23
Topics in Organometallic Chemistry

Topics in Organometallic Chemistry
Recently Published and Forthcoming Volumes

Regulated Systems for Multiphase Catalysis

Volume Editors: Walter Leitner · Markus Hölscher

With contributions by
A. Behr · P. Claus · E. Dinjus · B. Drießen-Hölscher · R. Fehrmann
J. A. Gladysz · J. Hasenjäger · M. Haumann · G. Henze · E. Hermanns
H. Hugl · C. Ionescu · J. Langanke · W. Leitner · P. Makarczyk
C. Maniut · M. Nobis · Y. Önal · F. Patcas · S. Pitter · A. Riisager
R. Roll · R. Schöbel · V. Tesevic · B. Turkowski · P. Wasserscheid

 Springer

The series *Topics in Organometallic Chemistry* presents critical overviews of research results in organometallic chemistry. As our understanding of organometallic structure, properties and mechanisms increases, new ways are opened for the design of organometallic compounds and reactions tailored to the needs of such diverse areas as organic synthesis, medical research, biology and materials science. Thus the scope of coverage includes a broad range of topics of pure and applied organometallic chemistry, where new breakthroughs are being achieved that are of significance to a larger scientific audience.

The individual volumes of *Topics in Organometallic Chemistry* are thematic. Review articles are generally invited by the volume editors.

In references *Topics in Organometallic Chemistry* is abbreviated *Top Organomet Chem* and is cited as a journal.

Springer WWW home page: springer.com
Visit the TOMC content at springerlink.com

ISBN 978-3-540-71074-5   e-ISBN 978-3-540-71076-9
DOI 10.1007/978-3-540-71076-9

Topics in Organometallic Chemistry ISSN 1436-6002

Library of Congress Control Number: 2008927187

Cover design: WMXDesign, Heidelberg
Typesetting and Production: le-tex publishing services oHG, Leipzig

Printed on acid-free paper

9 8 7 6 5 4 3 2 1 0

springer.com

Volume Editors

Prof. Dr. Walter Leitner

Institut für Technische
und Makromolekulare Chemie
RWTH Aachen
Worringerweg 1
52074 Aachen
Germany
leitner@itmc.rwth-aachen.de

Dr. Markus Hölscher

Institut für Technische
und Makromolekulare Chemie
RWTH Aachen
Worringerweg 1
52074 Aachen
Germany
hoelscher@itmc.rwth-aachen.de

Editorial Board

Dr. John M. Brown

Dyson Perrins Laboratory
South Parks Road
Oxford OX13QY
john.brown@chem.ox.ac.uk

Prof. Alois Fürstner

Max-Planck-Institut für Kohlenforschung
Kaiser-Wilhelm-Platz 1
45470 Mülheim an der Ruhr, Germany
fuerstner@mpi-muelheim.mpg.de

Prof. Peter Hofmann

Organisch-Chemisches Institut
Universität Heidelberg
Im Neuenheimer Feld 270
69120 Heidelberg, Germany
ph@uni-hd.de

Prof. Gerard van Koten

Department of Metal-Mediated Synthesis
Debye Research Institute
Utrecht University
Padualaan 8
3584 CA Utrecht, The Netherlands
g.vankoten@chem.un.nl

Prof. Manfred Reetz

Max-Planck-Institut für Kohlenforschung
Kaiser-Wilhelm-Platz 1
45470 Mülheim an der Ruhr, Germany
reetz@mpi.muelheim.mpg.de

Prof. Pierre H. Dixneuf

Campus de Beaulieu
Université de Rennes 1
Av. du Gl Leclerc
35042 Rennes Cedex, France
Pierre.Dixneuf@univ-rennesl.fr

Prof. Louis S. Hegedus

Department of Chemistry
Colorado State University
Fort Collins, Colorado 80523-1872
USA
hegedus@lamar.colostate.edu

Prof. Paul Knochel

Fachbereich Chemie
Ludwig-Maximilians-Universität
Butenandtstr. 5–13
Gebäude F
81377 München, Germany
knoch@cup.uni-muenchen.de

Prof. Shinji Murai

Faculty of Engineering
Department of Applied Chemistry
Osaka University
Yamadaoka 2-1, Suita-shi
Osaka 565
Japan
murai@chem.eng.osaka-u.ac.jp

Topics in Organometallic Chemistry
Also Available Electronically

For all customers who have a standing order to Topics in Organometallic Chemistry, we offer the electronic version via SpringerLink free of charge. Please contact your librarian who can receive a password or free access to the full articles by registering at:

springerlink.com

If you do not have a subscription, you can still view the tables of contents of the volumes and the abstract of each article by going to the SpringerLink Homepage, clicking on "Browse by Online Libraries", then "Chemical Sciences", and finally choose Topics in Organometallic Chemistry.

You will find information about the

– Editorial Board
– Aims and Scope
– Instructions for Authors
– Sample Contribution

at springer.com using the search function.

Color figures are published in full color within the electronic version on SpringerLink.

This volume is dedicated to the memory of
Prof. Dr. Birgit Drießen-Hölscher
(August 1, 1964 – November 16, 2004)

The Editors and Authors

Preface

Soluble organometallic catalysts are an inevitable tool in contemporary organic synthesis. Two Nobel Prizes in the first decade of the 21st century emphasise the continuing scientific progress and technical relevance of this principle. In fact, homogenous catalysts have long made their way into an increasing number of industrial applications on all scales, ranging from bulk chemicals to complex multi-step processes. Among the most prominent examples are the Monsanto and Cativa processes for the production of acetic acid, various processes for the hydroformylation of olefins to yield aldehydes and other oxo-products, the production of (–)-menthol at Takasago, and the synthesis of the blockbuster herbicide Metolachlor at Syngenta. These successful technical processes are, however, only the tip of an iceberg of potential applications involving the general strategy of homogeneous transition-metal catalysis. One of the major obstacles to an even broader use of homogeneous catalysts is the inherent difficulty to isolate the products from the reaction mixture containing the dissolved catalyst and to recycle the metal-containing component in active and selective form. The development of innovative and generic methodologies for catalyst immobilisation is therefore key to the future development of this important field of chemical science and technology.

The soluble nature of the active species is the basis for this success story as it allows modification and optimization of the catalytic performance on a molecular level. At the same time, it is also the main drawback of the approach, as products and catalysts are contained in the same phase at the end of the reaction. Efficient separation is often crucial, however, because the costly and highly specialized catalysts should be recovered and recycled for economic and environmental reasons. Furthermore, the specifications and applications do not allow even trace amounts of metals or other catalyst components in many products.

A very elegant solution to solve this problem is the introduction of either a permanent or a temporary phase boundary between the molecular catalyst and the product phase. The basic principle of multiphase catalysis has already found implementation on an industrial scale in the Shell higher olefin process (SHOP) and the Ruhrchemie/Rhône-Poulenc propene hydroformylation process. Over the years, the idea of phase-separable catalysis has inspired many chemists to design new families of ligands and to develop new separation and recycling strategies. The fundamental challenge in this field is to meet the right balance between sufficient contact of the substrates with the active species during the reaction step and distinct partitioning of catalyst components and products in the separation step. This can be influenced across all scales, by the molecular design of the catalyst, the choice of the reaction medium and solvent, and finally the reaction engineering implementation.

In 2002, the German network of competence for catalysis *ConNeCat* initiated a program through the Federal Ministry for Science and Education to establish selected Lighthouse Projects for emerging trends in catalysis research. One of these projects was devoted to tuneable systems for organometallic multiphase catalysis, aiming to develop innovative technologies for the immobilization of organometallic catalysts in solution. In particular, the project was dedicated to novel approaches using specially designed reaction media (smart solvents) and catalysts (smart ligands). The common key feature of the systems under investigation was the possibility to minimise mass transfer limitations during reaction and to optimise phase separation during work-up by control of process parameters, such as temperature, pressure, or composition. Seven academic groups formed the scientific consortium and four industrial partners covering the supply chain from bulk chemicals to specialized products supported and accompanied the research activities.

The present volume provides an overview of the different strategies followed within this network. The individual approaches are put into the context of the general development in the field and at the same time provide detailed information and hands-on experience with practical aspects of multiphase catalysis. The individual chapters address different and in many ways complementary approaches. The results and discussion emphasise that no single solution can give a universal answer to the problem of catalyst immobilization, but different strategies need to be developed for reaction types comprising substrates and products of various polarities and solubility properties. At the same time, general patterns emerge to provide guidelines of how to control the solubility and partitioning properties of catalytic systems for a given reaction/separation sequence.

The scientific adventure of this unique project was an exciting experience for us and we are most grateful to all fellow scientists and their motivated and enthusiastic group members that were part of the team. Their hard work and their achievements are cited and featured in this volume. In addition, we would like to thank Dr. Katja Stephan from Projektträger Jülich for accompanying the cluster throughout the program with a perfect balance between scientific freedom and administrative clarity. We also gratefully acknowledge the continuous support of Dr. Kurt Wagemann and Dr. Dana Demtröder at DECHEMA, which was vital for the formation, dissemination, and structural impact of the Lighthouse Project. In particular, the initiation of the biannual conference series on "Green Solvents" was an important stepping stone in this process. Finally, thanks are due to the Springer team for their help throughout the process and especially to Dr. Marion Hertel for providing the opportunity to assemble this volume.

Aachen, February 2008   Prof. Dr. Walter Leitner (Editor)
                                                    Dr. Markus Hölscher (Co-Editor)

Contents

Top Organomet Chem (2008) 23: 1–17
DOI 10.1007/3418_044
© Springer-Verlag Berlin Heidelberg
Published online: 8 July 2006

Multiphase Catalysis in Industry

Herbert Hugl[1] (✉) · Markus Nobis[2]

[1] Saltigo GmbH, 51369 Leverkusen, Germany
huglherbert@yahoo.de

[2] Symrise GmbH & Co. KG, Research & Development, Process Research,
  Mühlenfeldstrasse 1, 37603 Holzminden, Germany

Abstract During the past two decades a variety of transition-metal catalyzed reactions have been introduced in synthetic organic chemistry. Among the most popular examples are asymmetric oxidations and reductions and a variety of cross-coupling reactions to form C – C and C-heteroatom bonds. Examples are the Heck reaction, Suzuki–Kumada and Sonagashira coupling reactions and the aryl-aminations introduced by Buchwald and Hartwig.

These reactions are homogeneously catalyzed using a metal complex containing expensive metals and ligands many of which are difficult to synthesize. In most cases the catalysts are very efficient, therefore quite often only millimolar amounts or less are applied. This portfolio of new reactions was introduced into lab-scale synthesis within a few years of their discovery and is today frequently used for first syntheses of active pharmaceutical ingredients and other high-value fine chemicals. Protocols for the use of these reactions in very small-scale combinatorial synthesis have been developed and more and more hits and later on development products resulted from these efforts. Pharmaceutical products need, because of the intensive and time-consuming clinical development, up to ten years until the first industrial-scale production has to be scheduled. As part of the process development a method to fully separate the catalyst components from the products after the reaction has to be worked out. Only extremely small residues of metals or of ligands are tolerated in active pharmaceutical ingredients (APIs). The concentration of metals such as Pt, Pd, Ir, Rh, Ru or Os has been limited to 5 ppm by the recommendations of the European Agency for the Evaluation of Medicinal Products (EMEA) [1]. A variety of methods can be applied to fulfil these requirements (as exemplified for palladium in [2]).

Besides environmental and health criteria, cost considerations also motivate the development of processes which enable the separation and reuse of the catalyst complex after the reaction. Chiral ligands or ligands frequently used for cross-coupling reactions are difficult to synthesize and only some ten's of kilograms are needed even for the production of many tons of a final product. Both the demanding synthesis and the production in kg-labs or small pilot plants makes these ligands very expensive. Prices ranging from 2000 to 100 000 $/kg and more are quite common [3]. Both ecological and economical constraints force chemical and pharmaceutical industries to establish processes for the separation of the catalyst after the reaction to achieve a multiple use of the catalysts.

Phase separation during or after a reaction is a proven method in chemical industry for the extraction of a chemical compound. It has also been successfully introduced into large-scale catalysis processes [4, 5] mainly using systems with one aqueous and one organic phase. Typical catalysts for enantioselective or cross-coupling reactions normally are soluble in organic solvents only, just like the products of the reaction. Therefore, these proven systems cannot be applied in these cases. All these reasons have led to an increasing interest in new solvent systems for homogeneously catalyzed reactions: Supercritical fluids [6], ionic liquids [7], thermomorphic solvents [8, 9] and fluorous phases [10] are the most widely studied new "green" solvents that open a door for a wide variety of applications in organic chemistry. Each one seems to have a promising potential for industrial use. What is missing is the experience with their use, a better knowledge about scope and limitation of the systems and in particular a breakthrough in large-scale application. In the same way as a new catalyst system normally needs many years of lab experience until a first technical use arises, new solvents will also need this time period prior to their first use at the ton or multi-ton scale.

The ConNeCat (ConNeCat is the German Competence Network Catalysis; http://www. connecat.de) lighthouse project "Regulated Systems for Muliphase Catalysis—Smart Solvents/Smart Ligands" which was funded by the German Ministry of Research and Education (BMBF) provided the opportunity to gain important scientific and technical experience with the new solvent systems. Seven academic groups and four industrial partners got the chance to modify and validate these new solvent systems within core processes of the four companies. Basic aspects and new concepts could be worked out at the universities and research institutes. The industrial groups compared the new technologies with the state of the art and pointed towards the large-scale applicability. Several examples for the use of the new multiphase systems in homogeneous catalysis have been worked out. Each of the four systems investigated shows promise for industrial use, but —as could be expected—every single system has different applications and limitations. This review summarizes some of the main results of this research network together with selected parallel developments from the literature taking an industrial point of view.

1
Ionic Liquids

Room-temperature ionic liquids exhibit many properties that make them potentially attractive as solvents for catalysis [11]. They dissolve a wide range of organic, inorganic and organometallic compounds as well as gases, e.g., CO_2, HCl, NH_3, and to a much lower extent also H_2 and CO. Many ionic liquids show temperature operating ranges of more than 300 °C compared to 100 °C of water. Most of the known ionic liquids are thermally stable. In many cases it is possible to extract the reaction products from the ionic liquid with organic solvents or supercritical CO_2. Sometimes products can be distilled off. Hydrophobic ionic liquids are favorable in aqueous biphasic systems. Many homogeneous catalysis reactions have been demonstrated in ionic liquids [12, 13]. These new solvents also show possible advantages if used in biocatalysis [14]. They dissolve most of the potential substrates and products for biocatalytic reactions without the inhibiting effects on enzymes that are often encountered with organic solvents. Enzyme stability is often even better than in traditional media. Co-factors of enzymes can be regenerated more efficiently than in common solvents [15] and whole-cell biotransformations have been shown in biphasic ionic liquid/water systems [16]. Therefore, ionic liquids are valuable solvents for highly enantioselective reactions using chemo- as well as biocatalysis [17]. Finally, it should be mentioned, that the properties of ionic liquids can be tailored by combining appropriate anions and cations [18]. This enables a wide range of chemical properties.

Showing so much promise it is not surprising that ionic liquids are already used within large-scale industrial applications and that further industrial processes are in development. The Dimersol/Difasol process developed by the Institut Francais du Petrole uses an ionic liquid to dissolve the catalyst and to separate the catalyst phase from the product [19]. The products of the reaction—C 8 olefins—are not soluble in the ionic liquid and form a second phase that can be easily separated. The nickel catalyst dissolved in the ionic liquid can be recycled. In addition, the catalyst shows in the ionic liquid increased activity and better selectivity to the desired dimers rather than to the undesired higher oligomers.

A further large industrial process has been introduced by BASF: BASIL (biphasic acid scavenging ionic liquids) [20]. This process generates the ionic liquid in situ from methylimidazole to remove acidic hydrogen chloride formed during the synthesis of an alkoxyphenylphosphine. In this case the ionic liquid is used as the extraction phase. Compared with the HCl removal by adding a base such as triethylamine and filtering of the waste solid chloride salt the economics of the process have been improved significantly.

Degussa has announced plans to introduce a large-scale hydrosilylation process using an ionic liquid to separate the catalyst from the product [21–23].

Room-temperature ionic liquids have the clear potential to become frequently used solvents in chemical industry in addition to water and organic solvents. Numerous publications and the review articles cited demonstrate the broad variety of possible applications and tremendous range of possible combinations of cations and anions [24]. The environmental benefit of having essentially no vapor pressure is just one of the reasons. Very important is the opportunity to use ionic liquids in existing vessels and equipment. Further investment and engineering development are not crucial for their industrial use. Another advantage for multiphase catalysis applications is that no modifications at the catalyst system are necessary to gain solubility in the ionic liquid in favorable cases. Modifications of an already very expensive catalyst system should always be avoided. The cost of the modification easily overcompensates the benefit.

From the industrial point of view there are still open questions and challenges for further research:

- Cost and commercial availability: More and more lab suppliers and a few large-scale suppliers offer ionic liquids. For a more frequent use of these solvents the commercially available variety has to be increased and cost should be reduced. There is good reason that cost reductions will be possible in the near future because at least some of the ionic liquids will potentially find use in very large applications besides catalysis. They are discussed for fuel desulfurization, separations, liquefication, gasification and chemical modification of solid fuels, as electrolytes or in connection with synthesis and application of new materials. Also applications such as azeotrope-breaking liquids, thermal fluids or lubricants are under consideration. Because of economy of scale in combination with such applications, the price of the solvent will decrease significantly.
- Toxicology: up to now only rare and preliminary data concerning toxicology, biodegradability and the impact on aquatic ecosystems have been available [25, 26]. More data are urgently necessary to select and design those ionic liquids for industrial applications which show ecological and toxicological benefits.
- Purification of ionic liquids following single or multiple use. Side- or decomposition products of chemical reactions or catalysts will accumulate in the solvent. Cheap and reliable methods have to be developed for purification. Very probably there will not be a universal method. More likely for each type of process a specific method will be set up.
- Halogen content: If halogens in the anion are not crucial for specific reactions performed in the ionic liquid, they should be avoided. Moisture sensitivity, halogenide transfers, alcoholysis and toxic effects are often connected with halogen atoms in the molecule [27]. In addition, the hydrolysis products HCl or HF act corrosively. Within the project reported by Wasserscheid and coworkers they successfully developed ionic liquids with alkylsulfate groups as anions to overcome the halogen content. These new solvents show very favorable properties.

A major step towards applicability of multiphase catalysis in ionic liquids is the development of "Supported Ionic Liquid Phase (SLIP)"-catalysis by the Wasserscheid group [28, 29]. The SLIP concept enables quasi-heterogeneous catalysis in ionic liquids and opens the door to the production of basic chemicals.

In summary, it can be expected, that more and more processes using ionic liquids as the solvent will be introduced in chemical and pharmaceutical industries within a foreseeable future. The halogen-free solvents and the SLIP concept developed within the reported BMBF-funded project are major milestones for further successful applications of these new solvents.

2
Thermomorphic Multicomponent Solvent Systems

For a long time a proven concept for product separation following chemical reactions has been to cool down the homogeneous reaction mixture and allow the product to crystallize or solidify. Very often also the educts of the reaction are not soluble at lower temperature. Appropriate choice of the solvent for the reactions is the prerequisite for the success of that operation. Chemists developing procedures for production acquire a lot of expertise selecting the optimal solvent for a chemical reaction which should enable both optimal reaction conditions and the opportunity to separate the product in a clean form.

The idea to use solvent systems enabling homogeneous reaction conditions at elevated temperatures and liquid/liquid phase separation at lower—preferably room—temperature seems to be obvious. Nevertheless, it is only recently that thermomorphic solvent systems gain attention [30–33] for product separation or multiphase catalysis [34, 35]. The main reasons for the delayed engagement is that an efficient choice of a useful solvent system is not easy to achieve. There is also a lack of experience with thermomorphic systems in general. Reactions are optimized to be carried out in solvents having certain distinct solubility and polarity characteristics. A thermomorphic solvent system of choice will have to fulfill these requirements and to show the thermomorphic effect in addition.

On the other hand thermomorphic solvent systems can be used for industrial applications within existing equipment. In principal, it is also possible to use unmodified ligands and catalyst complexes for homogeneously catalyzed reactions. Therefore, the economic hurdles to use the system in industrial practice are not very high.

During the project Regulated Systems for Multiphase Catalysis—Smart Solvent/Smart Ligands researchers from universities closely collaborating with chemical companies had the opportunity to prove the concept of thermomorphic solvent systems by application to existing industrial processes. It could be shown that appropriate selection of a solvent system enables very high conversions and attractive selectivities, e.g., in telomerization and hydroformylation reactions. Most important is the elaboration of an expert system for the selection of a useful thermomorphic solvent system, e.g., for C – C-bond forming reactions using Pd-ligand complexes as the homogeneous catalyst. The loss of the expensive catalyst system was considerably reduced in most cases.

Thermomorphic solvents show promise for upcoming use in chemical industry. This is also the case for polymeric thermomorphic solvents [36]. But there are also limitations and items for further investigations:

- Often reactions only work satisfactory in special solvents. It is not very likely, that thermomorphic solvents or solvent mixtures will extend these options.
- Even considering that the recovery of an expensive catalyst system offers strong economic benefits, the option of a multiple use of a two- or three-component solvent system should be shown. Otherwise separation of the solvent mix into its components would compensate for potential benefits.
- Expert systems for the design of the solvent mix should be further developed. In particular, proposals including the type of reaction working in a particular system should be included. The availability of larger and more reliable data sets would reduce the hurdles for applications. Especially for active pharmaceutical and agrochemical ingredient process developments, which have to be done within a short time frame. There is no time available to adapt new solvent systems. Therefore, sufficient data should be available to select a solvent mix within a few experiments.
- In addition, advice for a later separation of the solvent mix would be helpful. Solvents have to be reused and purification of the solvents used is crucial.

Thermomorphic solvent systems are at a relatively young stage of development. Compared to ionic liquids or supercritical CO_2 there is much less experience available. Large-scale applications are unknown at present. There are a lot of options for the future but these will depend on further research in the area.

3
Supercritical Carbon Dioxide in Homogeneous Catalysis

Supercritical carbon dioxide ($scCO_2$) is a well-established solvent for applications in extraction processes. During the last 40 years, there has been an implementation of large-scale processes, e.g., the extraction of caffeine [6] and the isolation of hop extracts from raw plant material [37]. These examples show that the usage of this supercritical fluid ($p_c = 73.8$ bar, $T_c = 31.1$ °C) is a state of the art operation in process technology.

In recent years, attempts have been made to make use of the advantages of the supercritical carbon dioxide in chemical reactions. The first technical examples concerning the use of carbon dioxide in a "pilot-plant scale" chemical reaction were heterogeneous catalyzed hydrogenation and radical polymerization [38–42]. Meanwhile, hydrogenation reactions have been scaled up in a 1000 t/a commercial multipurpose plant.

The supercritical CO_2 was adopted in homogeneous catalysis due to its useful physicochemical properties towards a multiphase approach [43–45].

In laboratory-scale homogeneous catalysis applications, in the last decade further investigations have been carried out in which a less soluble organo-metallic catalyst system was utilized for metathesis reactions [46]. Under RCM-conditions, it was possible to convert substrates with functional groups that were problematic due to their potential to inactivate the ruthenium cata-lyst; here, the conversion in supercritical carbon dioxide avoids the protection of critical amino groups as an additional synthetic step. Consequently, it was possible to synthesize a number of carbo- and heterocyclic products with varying ring size (C_4 to C_{16}).

From an industrial point of view, homogeneous catalysis has significant advantages concerning selectivities and due to mild reaction conditions [47]. In fact, there is only a limited number of processes established in industrial applications because of the disadvantageous separability of the catalyst from substrate and product. A possible and convenient solution for this limitation can be the application of supercritical carbon dioxide as part of a reaction system due to the following:

- Good to excellent availability of carbon dioxide allows large-scale applica-tion of the agent.
- Inexpensive, decreases the investment and increases the economic effi-ciency of the whole process.
- Compressed CO_2 does not have any influence on the green-house effect, it is moreover not classified as a volatile organic chemical (VOC). Conse-quently, it may be regarded as a green solvent.
- The physical properties of the solvent can be varied by controlling the density in the supercritical area allowing adjustment of the reaction con-ditions for several chemical problems. Therefore, it will be possible to use supercritical carbon dioxide both as a catalyst phase and in other reactions as an extracting agent, which was shown earlier by carrying out catalytic hydrogenation, hydroformylation and $C - C$ coupling reac-tions [48–50] in this medium. Moreover, this methodology opens up the possibility of separating the catalyst from the product through switching of the conditions in the reactor [48, 51].
- Because of the physical properties and chemical behavior concerning cor-rosiveness and toxicity, it may be used in existing high-pressure plants and facilities at fine chemical industry production sites. This allows a fast and easy change to an environmentally benign solvent.
- In comparison to the classical applied multiphase homogeneous sys-tems, carbon dioxide differs beneficially because reactions can take place without the presence of any phase boundaries. Because of the physico-chemical properties, there is a reminiscence of both a gas and a liquid; the multiphase system is alleviated to a biphasic reaction mixture, in which

mass transfer limitation is reduced to only one phase boundary. This leads to advantages concerning conversions with gaseous reactants, e.g., hydrogenations and hydroformylation reactions. For industrial applications, this could lead to an increase in the space-time-velocity of a process. In these biphasic systems, the catalyst has to be immobilized in ionic liquids, polyether or water as a stationary phase, which are moreover able to stabilize the applied catalyst. To immobilize the catalyst, it is necessary to introduce functionalities, that are able to anchor the compound in the catalyst phase, e.g., polyether moieties or polar groups. Ionic catalysts can also lead to an immobilization in suitable media, like aqueous solutions or ionic liquids [52].

In addition to the advantages listed, there are also some disadvantages that arise from the use of carbon dioxide as a reaction medium in homogeneous catalysis:

- For systems where the catalyst is required in the CO_2 phase a modification of the ligand periphery to increase the solubility in the supercritical medium is usually necessary. This has to be mostly done via introduction of perfluorinated tags ("ponytails") which causes expensive and/or sometimes difficult synthetic operations.
- The solubility of the organic substrates/products can also be a limiting factor, as large amounts of gas must be compressed and expanded.

One essential aspect of the publicly funded project "Smart Solvents/Smart Ligands" was the development of new and useable techniques for the isolation and reuse of organometallic catalysts. The work of Leitner and Dinjus shows various possibilities for the immobilization of modified and conventional catalytic systems.

Concerning supercritical carbon dioxide, the investigations of Leitners group allows us to imagine the wide range of uses for this solvent, both in the separation and catalyst phase in different applications, e.g., in hydroformylation reactions using polyether-modified ligand systems, in which the CO_2 may act as an extracting agent for the generated product ("phase switch"; the CO_2 is added after conversion of the alkenes to the aldehydes; the insoluble catalyst is left in the reaction vessel and may be reused) [53, 54].

From an industrial point of view, it is remarkable that the introduction of the polyether moiety allows the separation of the products and the reuse of the catalyst without any loss of activity and selectivity. For the development of technical processes, this aspect may lead to a facile downstream processing of the crude product.

Moreover, it could be demonstrated that by applying this separation methodology the same catalyst is useable for different types of reactions in a consecutive manner [54]. For fine chemical applications, this may lead to a high flexibility of catalytic processes and decrease their operational costs because of their good adaptability.

An "inverted" system with the catalyst in the CO_2 phase and the products in an aqueous phase was used to convert very polar substrates e.g., itaconic acid via hydrogenation [55, 56]; in this case the carbon dioxide acted as a solvent for the CO_2-philic catalyst comparable to a few other mentioned systems in the literature [57, 58].

Because of the mentioned disadvantages of modified catalytic systems, there is an aim to apply well-known conventional, unmodified ligands for the development of organometallic catalytically active compounds. Therefore, it is necessary to gain deep insight into the physical and solubility properties both of the ligands and of the organometallic compounds in scCO₂.

The detailed investigations done by Dinjus and coworkers in the area of industrially important types of reactions (e.g., hydroformylation) concerning kinetic data and physicochemical properties of the applied systems showed the potential to achieve the desired reaction and/or separation conditions via variation of the physical conditions in the reaction vessel [59].

Establishing this reaction protocol, similar to the immobilized homogeneous systems mentioned earlier, the unmodified systems do have the potential to provide access to industrial applications, for economic reasons, and because of the good availability of the catalyst precursors; the systems used have to be validated further concerning two crucial aspects:

• Leaching of metal into the product/contamination of the products.

The loss of the metal into the crude separated product can lead to difficulties in further stages of the chemical process (purification of the crude material, following reaction steps, so-called "downstream processing").

• Loss of ligands/destabilization of the catalyst.

In most cases, the stability of the employed catalytic system is crucial for the stability of a catalytic process. In homogeneous catalyzed reactions, the stability of the used organometallic compound is determined by the partial loss of the ligands during the process. Moreover, through the loss of the coordinated ligands, a change of catalytic performance can be assumed.

Also, in several catalytic processes, the loss of the ligand during the reaction leads to an increase in operating costs of the process, due to the fact that the expense of the ligand is in many cases much higher than that of the chosen metal. The loss of the phosphorus-containing ligand is also crucial for the implementation of the homogeneous catalyzed process in fine chemical production; the content of these substances, e.g., in cosmetic materials, is critical due to their irritating and allergenic potential [60]. The recent research work of the Dinjus group concerning the recovery and reuse of the catalytically active compounds together with the results of earlier investigations represent a starting point with the potential to implement systems based on supercritical fluids for homogeneous catalysis with an unmodified substitution pattern of the catalysts.

The further optimization and development concerning stability and selectivity of the organometallic catalyst in these kinds of media and the application of isolation methodologies similar to CESS (catalysis and extraction using supercritical solutions [43]) together with the physical and chemical advantages of supercritical fluids can lead to high potential catalyst matrices that fulfil the requirements of industrial processes both for bulk and fine chemicals.

In future research, it may be interesting to combine supported liquid phase (SLP) catalysts or supported aqueous phase (SAP) catalysts [61, 62] with scCO$_2$ as a mobile phase.

In particular the recent investigation as part of the project Smart Solvents/Smart Ligands of the SLPC with PEG and ionic liquids ("SILP") as the catalyst carrier opened up new possibilities for the immobilization of homogeneous catalysts. By increasing the stability of the catalytic performance, this concept may have the potential to be kept in mind for industrial catalysis in combination with environmentally benign supercritical or compressed carbon dioxide. In addition, this methodology is able to provide access to some chemical applications and processes because of the easy and facile preparation of the coated materials.

4
Perfluorinated Thermomorphic Catalysts

Another approach to isolate the catalyst from the products is the application of perfluorinated catalytic systems, dissolved in fluorinated media [63], which are not non-miscible with the products and some commonly used solvents for catalysis like THF or toluene at ambient temperature. Typical fluorinated media include perfluorinated alkanes, trialkylamines and dialkylethers. These systems are able to switch their solubility properties for organic and organometallic compounds based on changes of the solvation ability of the solvent by moving to higher temperatures. This behavior is similar to the above-mentioned thermomorphic multiphasic PEG-modified systems [65–67].

Potential advantages of perfluorinated solvents are their inertness against most of the employed reaction conditions, their low acute toxicity and a high solubility for gases making it interesting to evaluate their introduction in homogeneous catalytic multiphasic protocols.

The developments of the fluorous biphasic systems was initiated by the work of Horvath and Rabai, who tried to overcome the disadvantages of the conventional homogeneous biphasic aqueous systems in the hydroformylation of higher alkenes concerning their low solubility in the aqueous catalyst phase [68, 69]. Another motivation for the study of this methodology is to avoid the isolation of the high-boiling products by distillation, which stresses both catalyst and product. Applying a fluorous biphasic catalytic system, the hydroformylation could be repeated several times without loss of activity and selectivity; the detected rhodium-leaching was about 1 ppm per mole aldehyde. Other examples for conversions in perfluorinated systems are hydrogenation, hydroboration and hydrosilylation reactions with a modified Wilkinson-type perfluorinated catalyst or the development of perfluoro-modified catalysts for the oligomerization of ethylene [70, 71].

The main limiting factors for catalytic conversions in perflourinated solvent systems on a bigger scale are given below:

- Evolution of harmful and corrosive decomposition products in the case of high temperatures.
- Ecotoxicology: The good inertness of the perflourinated systems against most reaction conditions will lead to a critical accumulation of non-biodegradable substances in the environment [72]. This waste deposition may lead to ecological problems in the future.
- From an economic point of view, conversions in perfluorinated solvents will increase the operational costs of the process, due to the price of the matrix.
- Availability: The solvents are often available in relatively small amounts only. A multi-kilogram synthesis in perfluorinated reaction media would be limited by the availability of the solvent. Moreover, similar to the development of perfluorinated, modified ligands for catalysis in scCO$_2$, the availability of the ligands can be limited by transfer from the laboratory synthesis to a technical scale. Typically, the degree of fluorination in a fluorous biphasic system is even considerably higher than for scCO$_2$ applications [73].

These critical aspects of the "classical" fluorous biphasic catalysis led in recent works to the development of protocols for the conversions with modified catalyst systems in non-fluorinated hydrocarbons as solvents. As part of the BMBF lighthouse project, Gladyzs and coworkers applied this concept to C – C coupling reactions (Suzuki reaction) and other metal-catalyzed addition reactions (hydrosilylation, selective alcoholysis of alkynes), which have direct relevance for the synthesis of fine chemicals and specialties [74].

An important step towards a possible application of these compounds in technical syntheses of chemicals was the successful demonstration of a thermomorphic reversible immobilization of perfluorinated catalysts on teflon or other solid fluorous matrices, which might be practiced in industrial low-scale applications, e.g., of pharmaceutical intermediates in the case of quantitative recovery of the organometallic compound. The facile separation due to their physicochemical behavior and the constant good performance in coupling reactions of the involved perfluorinated pincer complex makes this system attractive for further investigations.

This development towards an ecologically and–from an industrial point of view—economically less critical catalytic system based on thermomorphic fluorous catalysts broadens the toolbox of the industrial research chemist and should be taken into consideration in future developments of chemical processes.

5
Reaction Engineering and Reactor Design for Multiphasic Reaction Systems

Today, for chemists in research and development, homogeneous catalysis is part of the synthetic toolbox, because of the earlier mentioned advantage of high selectivities and mild reaction conditions. Since the establishment of homogeneous catalyzed processes in the chemical industry, the aim has been to develop techniques which facilitate the separation and the reuse of the organometallic catalyst systems. In addition, for the development of new molecularly designed or solvent-based separation concepts reaction engineering aspects of multiphase systems are crucial for their implementation in industrial processes. The most successful large-scale application of multiphase catalysis is based on the development of the trisulfonated triphenylphosphine (TPPTS) by Kuntz and coworkers with which catalyst recovery and recycling can be realized through aqueous biphasic catalysis [75]. By using aqueous biphasic catalysis in the hydroformylation of short alkenes, the loss of the precious metal into the product can be suppressed to 10^{-9} kg Rh/kg product. Moreover, in some applications, further investigations have shown the option to minimize consecutive reactions. Examples are given by Mortreux (Ru-catalyzed biphasic hydrogenation of benzene to cyclohexene) [62] and Driessen-Hölscher (Pd-catalyzed, selective telomerization of butadiene and ammonia to primary allylic octadienylamines) [76, 77], which were realized on a laboratory or pilot-plant scale.

In the case of aqueous multiphasic catalytic conversions, the reaction rate can be strongly affected by the ability of the substrate to move over into the catalyst phase. For biphasic hydroformylation, the velocity decreases with increasing chain length of the olefins due to their lower solubility in the aqueous phase [78].

For the design and engineering of technical biphasic processes, it is necessary to gain insight into crucial physical and chemical reaction parameters [79]:

- Batch operation: For the design of batch reactors for biphasic conversion the type of stirring device is an essential aspect to generate a narrow distribution with small droplet sizes which is equivalent to high surfaces [36]. Together with the diffusion ability (diffusion coefficient) of the used solvent system, the generated surface can be correlated directly to the rate of the conversion.

- Continuous operation: In flow-type reactors, e.g., loop reactors, the space velocity of the reaction is determined through the installed static mixing device that is used to generate the dispersion, together with the velocity of the circulating medium (catalyst- and substrate/product phase). Knowledge of these parameters allows one to set up a kinetic model for the investigated reaction.

- Process conditions: The optimization of the process conditions such as temperature, mixing conditions, correct choice of both extraction and catalyst phase, uptake of gaseous reactants (hydrogen, syngas) into the catalyst phase are of essential importance to develop a high-yielding and selective synthetic route to the desired product in the biphasic system.

- Cocatalysts and promotors: Cocatalysts and promotors are able to activate the reactants and to increase the velocity of the reaction. One example is given by the hydrodimerization of butadiene, in which the presence of carbon dioxide and a tertiary amine increases the nucleophilicity of the water. Without cocatalysts or promotors, the conversion does not take place due to critical environmental changes in the reaction medium. In some catalytic conversions, the reaction can only proceed by addition of these compounds, which interfere at an earlier (telomerizations) or later (e.g., C – C coupling reactions) stage of the catalytic cycle. The use of surfactants to inverse the interfacial exchange in aqueous biphasic systems can also be applied to influence reaction rates.

The investigations done by Claus and coworkers as part of the project Smart Solvents/Smart Ligands focussed on the selective hydrogenation of α,β-unsaturated carbonyl compounds and showed the potential of aqueous multiphasic catalysis for the production of chemicals for fine chemistry, e.g., fragrance materials.

The first results of the batch hydrogenation of prenal and citral to geraniol and nerol provided evidence for the use of aqueous biphasic catalysis to increase the selectivity of the conversion of the substrates; the accumulation of byproducts can be nearly suppressed by the fast extraction of the product from the catalyst phase. For this reason, the distribution of the product between extraction and catalyst phase had been studied in detail.

Further investigations on the selective hydrogenation of α,β-unsaturated aldehydes showed the general use of the applied concept. The hydrogenation of four different unsaturated aldehydes to their corresponding allylic alcohols could be conducted with high conversions and selectivity. As a result of these research activities, one is now able to set up the best reaction system for the biphasic selective conversion of unsaturated aldehydes.

The development of a kinetic model for the biphasic reaction broadens the knowledge about these types of conversions and may help to obtain more detailed information about the influence of the multi parameter system involved.

As the next step in multiphasic hydrogenation, the design and implementation of a continuously driven loop reactor as a laboratory-scale plant model led to comparable selectivity applying the same water soluble ruthenium-based catalyst system.

Further investigations and optimization generated important and essential data for the development of other multiphasic reaction systems. The detailed data concerning reactor and static mixer design is of significant interest for the research and development of continuous multiphasic processes both in academic and industrial research departments. A good agreement between experimental data and computational data was also shown during the project.

6
Summary and Outlook

As a result of the joint research within this project network concerning multiphase catalysis in smart solvent systems, a basic feasibility for industrial applications of the new systems has been widely proven. Multiphase catalysis in supercritical fluids and ionic liquids has already gained large-scale application in the chemical industry. The results of the project widen the options for their use in the future significantly. Thermomorphic systems show good potential for industrial use including large-scale production. The examples studied and the expert system developed within the project are large steps towards extended use. But the systems are still young and further research in the area will be necessary. Fluorous phases show promise mainly for use within laboratory or small-scale operations. They will enable the recovery of toxic metals as well as expensive ligands. This will be very beneficial but for large-scale chemical production flourous systems will probably remain too expensive within the near future.

Over the three year project very valuable information has been gained and significant progress has been made using industrially relevant prototypes for the new techniques. The interaction between the academic research groups and the industrial partners has been very inspiring for both parts and will contribute significantly to future developments and implementations of the investigated systems.

In this volume of *Topics in Organometallic Chemistry* a concise summary of the developments within the network is put in perspective of the recent advancements within the field in general. It is hoped that this will encourage further applicants to use multiphase catalysis techniques in fine chemical production. In addition, students and post doctoral fellows will transfer their expertise obtained within the project groups to their future employers. Therefore, the results summarized in this volume have contributed to foster the strong position of German chemistry in the area of multiphase catalysis, and provide a good basis for further economic success.

References

1. Note for Guidance on Specification Limits for Residues of Metal Catalysts, Evaluation of Medicines for Human Use (2002) The European Agency for the Evaluation of Medicinal Products, London, http://www.emea.en.int
2. Christine E, Garrett H (2004) Adv Synth Catal 346:889
3. Blaser HU (2006) In: de Vries J, Cornelis J (eds) The Handbook of Homogeneous Hydrogenation. Wiley, Weinheim
4. Cornils B, Herman WA (eds) (2002) Applied Homogeneous Catalysis with Organometallic Compounds, 2nd edition. Wiley, Weinheim
5. Vogt D, Horath J, Olivier-Bourbigou H, Leitner W, Mecking S (2005) In: Cornils B, Hermann WA (eds) Multiphase Homogeneous Catalysis. Wiley, Weinheim
6. Leitner W (2002) Accounts of Chemical Research 35:746
7. Wasserscheid P, Welton T (eds) (2003) Ionic Liquids in Synthesis. Wiley, Weinheim
8. Bergbreiter DE (2002) Chem Rev 102:3345
9. Jin Z, Wang Y, Zheng X (2004) In: Cornils B, Hermann WA (eds) Aqueous-Phase Organometallic Catalysis. 2nd edition. Wiley, Weinheim, p 301
10. Curran DP (1998) Angew Chem Int Ed Engl 37:1174
11. Sheldon R (2001) Chem Commun, p 2399
12. Holbrey JD, Seddon KR (1999) Clean Products and Processes. Springer, Berlin Heidelberg New York, 1:223–236
13. Holbrey JD (2004) Chem Today 22:35
14. van Rantwijk F, Lan RM, Sheldon RA (2003) Trends Biotechnol 21:131
15. Eckstein M, Villeta Filko M, Liese M, Kragl U (2004) Chem Commun, p 1084
16. Pfründer H, Amidjojo M, Kragl U, Wenster-Bolz D (2004) Angew Chem Int Ed 43:4529
17. Song C (2004) Chem Commun, p 1033
18. Forsyth StA, Pringle JM, MacFarlane DR (2004) Anst J Chem 57:113
19. Chauvin Y, Gilbert B, Giubard I (1990) Chem Commun, p 1715
20. Seddon KR (2003) Nat Mat 2:363
21. Goldschmidt AG, EP 1 382 630 A1
22. Weyershausen B, Hell K, Hesse U (2003) ACS National Meeting, Sept. 8, New York
23. Weyershausen B, Wehmann K (2004) Green Solvents for Synthesis. Bruchsal, Germany
24. Holbrey JD, Seddon KR (1999) Clean Technol Environ Pol, p 223
25. Wagner M (2005) Chem Today 23:24
26. Jarstorff Bl, Stoermann R, Ranke J, Moelter K, Stock F, Oberheitermann B, Hoffmann W, Hoffmann J (2003) Green Chem 5:136
27. Wasserscheid P, Maase M (2003) Chem Ing Tech 75:1150
28. Riisager A, Fehrmann R, Flicker S, van Hal R, Haumann M, Wasserscheid P (2005) Angew Chem Int Ed 44:815
29. Mehnert CP (2005) Chem Eur J 11:14
30. Bergbreiter DE (2002) Chem Rev 102:3345
31. Bergbreiter DE, Chandran R (1987) J Am Chem Soc 109:174
32. Bergbreiter DE, Zhang L, Mariagnanam VM (1983) J Am Chem Soc 115:9295

33. Karakhanor EA, Runova EA, Berezkin GV, Meimerovets EB (1994) Macromol Symp 80:231
34. Jin Z, Zheng X (1996) Thermoregulated Phase-transfer Catalysis. In: Cornils B, Herrmann WA (eds) Aqueous-Phase Organometallic Catalysis, Chap. 4, Sect. 6.3. Wiley, Weinheim, p 233
35. Behr A, Fängewisch C (2001) Chem Ing Tech 73:874
36. Bergbreiter DE (2001) J Polm Science Polym Chem Ed 39:2352
37. Zosel K (1978) Angew Chem Int Ed Engl 17:702
38. McCoy M (1999) Chem Eng News 77:11
39. Pickel KH, Steiner K (1994) Proc 3rd Int Symp Supercritical Fluids, Strasbourg, France, p 25
40. Freemantle M (2001) Chem Eng News 79:30
41. Devetta L (1999) Catal Today 48:337
42. Licence P, Ke J, Sokolova M, Ross SK, Poliakoff M (2003) Green Chem 5:99
43. Koch D, Leitner W (1998) J Am Chem Soc 120:13398
44. Jessop PG, Ikarya T, Noyori R (1995) Chem Rev 95:259
45. Cole-Hamilton DJ (2003) Science 299:1702
46. Fürstner A, Leitner W, Koch D (1997) Angew Chem Int Ed Engl 36:2466
47. Cornils B, Herrmann WA (1998) Aqueous Organometallic Catalysis. Wiley, Weinheim
48. Leitner W (1999) In: Knochel P (ed) Reactions in Supercritical Carbon Dioxide (scCO2) in Modern Solvent Systems. Top Curr Chem 206:107
49. Morita DK, David SK (1998) Chem Commun, p 1397
50. Wegner A, Leitner W (1999) Chem Commun, p 1583
51. Sellin M, Cole-Hamilton DJ (2000) J Chem Soc, Dalton Trans, p 1681
52. Solinas M, Pfaltz A, Leitner W (2004) J Am Chem Soc 126:16124
53. Leitner W, Scurto AM (1998) Imobilization of Organometallic Catalysts using Supercritical Fluids. In: Cornils B, Herrmann WA (eds) Aqueous Organometallic Catalysis. Wiley, Weinheim, p 664
54. Solinas M, Leitner W (2005) Angew Chem Int Ed 44:1346
55. Mc Carthy M, Stemmer H, Leitner W (2002) Green Chem 4:501
56. Burgemeister K, Franciò G, Hugl H, Leitner W (2005) Chem Commun, p 6026
57. Verspui G, Elbertse G (2000) Chem Commun, p 1363
58. Verspui G, Papadogianakis (2001) J Organomet Chem 621:337
59. Dahmen N, Griesheimer P, Makarczyk P, Pitter S (2005) J Organomet Chem 690:1467
60. Merk HF (2002) Zeitschrift für Hautkrankheiten 77:466
61. Chan WC, Lau CP (1994) J Organomet Chem 464:103
62. Bricout H, Mortreux A (1998) J Organomet Chem 553:469
63. Hildebrandt J, Prausnitz JM (1970) In: Van Nostrand R (ed) Regular and Related Solutions, Chap 10. New York
64. Horvath I (1998) Flourous Phases. In: Cornils B, Herrmann WA (eds) Aqueous Organometallic Catalysis. Wiley, Weinheim, p 549
65. Fell B, Jin Z (1997) J Mol Catal A 116:55
66. Jin Z, Fell B (1996) J Prakt Chem 338:124
67. Bergbreiter DE (2000) J Am Chem Soc 122:9058
68. Horvath I, Rabai J (1994) Science 266:72
69. Horvath I (1995) EP 633062 A1
70. Horvath I, Gladysz JA (1996) 10th Int Symp on Homogeneous Catalysis, Princton, USA, Abstract p A59

71. Vogt M (1991) PhD thesis, Rheinisch Westfälische Technische Hochschule Aachen, Germany
72. Schwaab K (2004) Bericht des Bundesumweltamtes, p 175
73. Leitner W (2003) In: Desimone JM, Tumas W (eds) Green Chemistry using Liquid and Supercritical Carbon Dioxide. Oxford University Press, Oxford, pp 81–102
74. Herrmann WA, Reisinger CP (1998) C – C-Coupling by Heck-type-Reactions. In: Cornils B, Herrmann WA (eds) Aqueous Organometallic Catalysis. Wiley, Weinheim, p 383
75. Kuntz EG (1987) Chemtech, p 570
76. Prinz T, Driessen-Hölscher B (1999) Chem Eur J
77. Prinz T, Keim W, Driessen-Hölscher B (1996) Angew Chem 108:1835
78. Baerns M, Hoffmann H, Renken A (2002) Chemische Reaktionstechnik. Wiley, Weinheim, p 67
79. Horvath I (1990) Catal Lett 6:43

Top Organomet Chem (2008) 23: 19–52
DOI 10.1007/3418_2006_057
© Springer-Verlag Berlin Heidelberg
Published online: 21 November 2006

Multiphase Catalysis in Temperature-Dependent Multi-Component Solvent (TMS) Systems

Arno Behr (✉) · Barbara Turkowski · Reina Roll · René Schöbel ·
Guido Henze

Universität Dortmund, Fachbereich Bio- und Chemieingenieurwesen,
Lehrstuhl für Technische Chemie A, Emil-Figge-Straße 66, 44227 Dortmund, Germany
behr@bci.uni-dortmund.de

Abstract The use of temperature-dependent multi-component solvent-systems (TMS) as a new recycling concept was investigated. The temperature dependency of the solvent systems and suitable compositions for various reactions were determined by cloud titrations. The results were summarized in an expert system for the solvent selection. Appropriate thermomorphic solvent systems were applied to different C – C bond-forming reactions: the telomerization of butadiene with ethylene glycol or with carbon dioxide, the isomerizing hydroformylation of *trans*-4-octene and the hydroaminomethylation of 1-octene with morpholine. In further investigations the carboxytelomerization and the synthesis of 4-nitrodiphenylamine were examined. Especially, if the polarity of the reaction mixture remains constant, high conversions of the substrates and high selectivities can be achieved. Thus, for the hydroformylation the conversion reaches a level of 99% and a selectivity to *n*-nonanal of about 80% in the TMS system propylene carbonate (PC)/dodecane/*N*-methylpyrrolidone (NMP). Similar results are obtained for the hydroaminomethylation: if *N*-octylpyrrolidone is used as a mediator the conversion of 1-octene and the selectivity of the corresponding amines reach 92%. For both reactions the catalyst can be easily recovered by a simple phase separation with only marginal loss of rhodium catalyst.

Keywords Catalyst recycling · Homogeneous catalysis · Temperature-dependent solvent systems

Abbreviations

acac	acetylacetonate
ad	adamantyl
bis-mes-IM	1,3-bis(mesitylene)imidazolium chloride
cod	cyclooctadiene
CST	critical solution temperature
DMF	dimethylformamide
DMSO	dimethyl sulfoxide
dvds	1,3-divinyltetramethyldisiloxane
EC	ethylene carbonate
HSP	Hansen solubility parameter
ICP-OES	inductively coupled plasma optical emission spectroscopy
IPA	isopropyl alcohol
MCH	methylcyclohexyl
MDPP	methyl polyethylene glycol diphenylphosphine
MA	maleic anhydride
NBP	N-benzylpyrrolidone
NCP	N-cyclohexylpyrrolidone
NEP	N-ethylpyrrolidone
NMP	N-methylpyrrolidone
NOP	N-octylpyrrolidone
OAc	acetate
PC	propylene carbonate
PEG	polyethylene glycol
PETPPO	polyethylene glycol triphenylphosphine oxide
PF	perfluoro-prod., product
PVC	polyvinyl chloride
rpm	rotations per minute
Rf$_x$	$- C_x F_{2x+1}$
THF	tetrahydrofuran
TMS system	temperature dependent or thermomorphic multi-component solvent system
TPPMS	triphenylphosphine-3-sulfonic acid lithium salt
TPPTS	triphenylphosphine-3,3′,3″-trisulfonic acid trisodium salt

1
Introduction

One of the most important difficulties in the use of homogeneous catalysis for technical applications is the recovery and reuse of the expensive metal catalyst. Various concepts for catalyst recycling are applied to overcome this problem in current technical applications. A simple method is to perform the catalytic reaction in a single phase and to separate the product and the catalyst afterwards by distillation of the product, as realized, for example, in the Wacker–Hoechst process (the palladium/copper catalyzed oxidation of ethylene to acetaldehyde) or in the Monsanto/Cativa processes (the rhodium/iridium-catalyzed carbonylation of methanol to acetic acid) [1]. The use of multi-phase systems, where the catalyst is dissolved in one phase and the product is located in the other phase is an alternative recycling concept. Technical applications for this principle are the Ruhrchemie/Rhône-Poulenc process and the production of α-olefins via the SHOP-process [1]. However, this concept is limited to processes where the organic reactants are sufficiently soluble in the polar catalyst phase. Otherwise mass transport limitations lead to a decrease of the reaction rate and to very low yields.

In the present paper we describe our recent development of a new recycling concept, which combines the advantages of a reaction in single-phase systems by overcoming typical mass transport limitations with the advantages of a convenient catalyst recycling in two-phase systems [2–5]. The use of temperature-dependent or thermomorphic multi-component solvent systems (TMS systems), that can be changed from two-phase to single-phase systems by simply raising the temperature, allows us to perform a reaction in a homogeneous phase followed by a phase split at a lower temperature. The TMS systems usually consist of three liquid components: a polar (s1), a non-polar (s2) and a semi-polar (s3) solvent or reactant. S1 and s2 are not or barely miscible with each other. In one of these components the catalyst is dissolved, the other one serves as an extraction agent for the reaction products. The semi-polar component s3 operates as a mediator for the other two components (Fig. 1). The miscibility gap between s1 and s2 is temperature-dependent and decreases usually with rising temperature. Depending on the composition and the temperature a mixture of s1, s2 and s3 is either homogeneous or heterogeneous. In the phase diagram the operating point describes the composition of the solvent system appropriate for the reaction. The reaction can

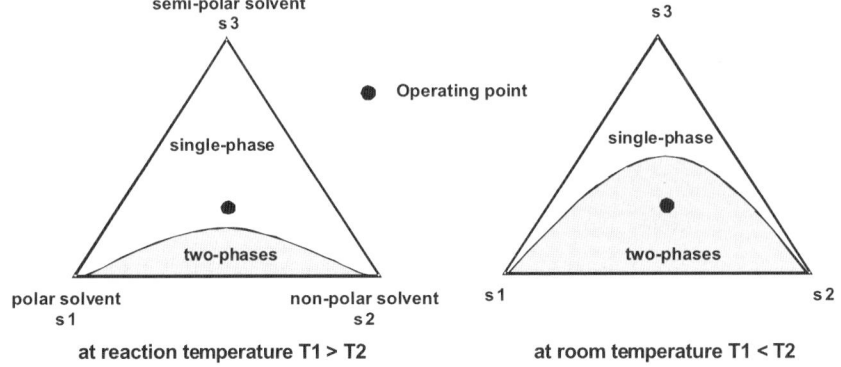

Fig. 1 Principle of temperature-dependent multi-component solvent systems (TMS systems)

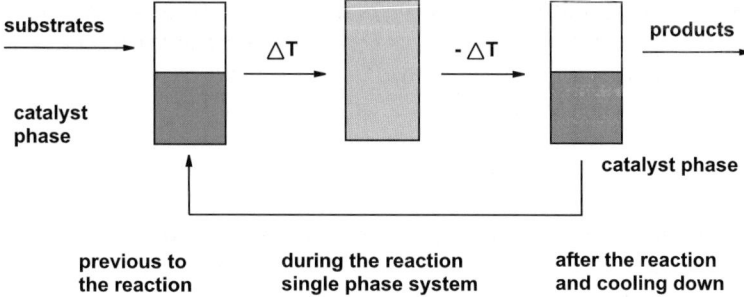

Fig. 2 Phase behaviour of a TMS system

be carried out in one homogeneous phase and after cooling the product mixture the catalyst can be easily recovered by a simple phase separation (Fig. 2).

The new recycling concept was applied to several C – C bond-forming reactions, for example, to the telomerization of butadiene with ethylene glycol or carbon dioxide, to the isomerizing hydroformylation of *trans*-4-octene and to the hydroamino-methylation of 1-octene with morpholine.

2
Homogeneous Catalysis in Thermomorphic Solvent Systems

2.1
Telomerization

2.1.1
Telomerization of Butadiene with Ethylene Glycol

The telomerization of butadiene with ethylene glycol was chosen as an example for a reaction of a polar and a non-polar substrate to a semipolar product (Scheme 1).

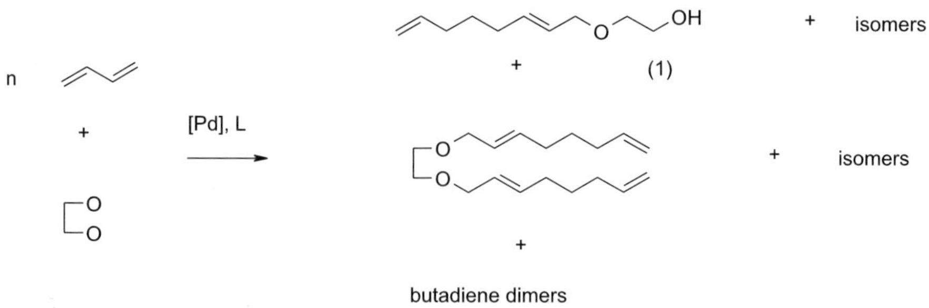

Scheme 1 Telomerization of butadiene with ethylene glycol

The linear monotelomer 2-(2,7-octadienyloxy)ethanol (1) can be hydrogenated to its saturated derivative 2-octyloxy-ethanol, which can be used as plasticizer alcohols for polymers like PVC. Some of our results for the telomerization of ethylene glycol in a biphasic aqueous system have already been published [6-8].

2.1.1.1
Temperature-Dependent Solvent Systems for the Telomerization Process

On the basis of the biphasic telomerization of butadiene and ethylene glycol in water, an ethylene glycol/water mixture in a weight ratio 2 : 1 was used as polar solvent s1 in the TMS system [9]. This ratio represents a typical composition of the reaction mixture at the beginning of the reaction. Toluene was applied as non-polar solvent s2 and several potential mediators s3 were tested. The polarity of s3 has to be intermediate between the polarities of s1 and s2. Adequate solvents were chosen on the basis of the polarity parameters of Kosower [10]. According to these parameters the aliphatic alcohols (C1 to C4), cyclic ethers like 1.4-dioxane and tetrahydrofuran (THF), dimethylformamide (DMF), dimethyl sulfoxide (DMSO), acetone and acetonitrile should be applicable to the TMS systems. Also some less-typical solvents for catalysis such as propylene carbonate (PC), polyethylene glycol 400 (PEG 400) and butyrolactone were used for the investigations. Two aspects have to be considered to decide if a solvent is suitable in a TMS system: the amount required to form a homogeneous solution and the temperature dependency of the solvent system. The latter has to be sensitive enough to obtain a biphasic system after the reaction by cooling of the product mixture. To avoid dilution effects and possible disadvantages connected to it (e.g. deceleration of the reaction, higher amounts of catalyst, larger reactor volume) the amount of mediator should not be kept to a minimum.

To determine the optimum composition of the solvent system the ethylene glycol/water mixture (s1) and toluene (s2) were mixed in a weight ratio of 1 : 3 at 80 °C and the mediator s3 was added dropwise until a homogeneous solution was formed. These cloud titrations were repeated at 60 °C, 40 °C and 25 °C. The required amount to obtain one single phase increases in the following order: isopropyl alcohol < *tert*-butanol < DMF < DMSO < acetone < butyrolactone < PEG 400 < 1.4-dioxane < THF < propylene carbonate (Fig. 3). The temperature dependency is also determined by the mediator s3 used. Solvent systems with alcohols isopropyl alcohol and *tert*-butanol behave almost independent of the applied temperature whereas systems with propylene carbonate or acetonitrile show the largest temperature dependency (Figs. 3 und 4). The following order was determined for the temperature dependency: *tert*-butanol < isopropyl alcohol < PEG 400 < THF < 1.4-dioxane, acetone, DMF < butyrolactone, DMSO < acetonitrile < propylene carbonate. From these data it can be concluded that the most appro-

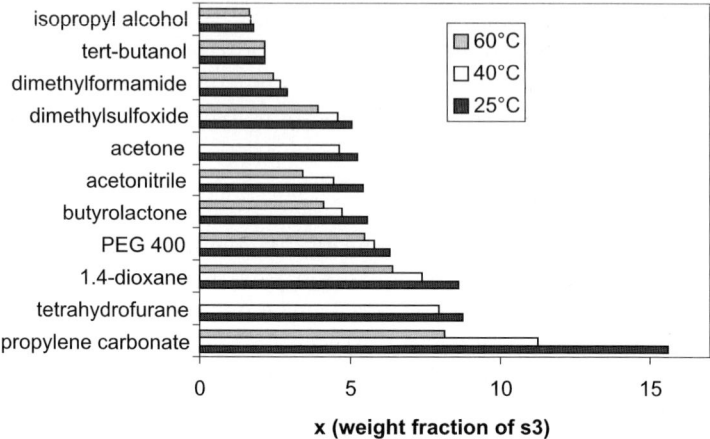

Fig. 3 Required weight ratio (ethylene glycol/water) : toluene : s3 = 1 : 3 : x for different mediators s3

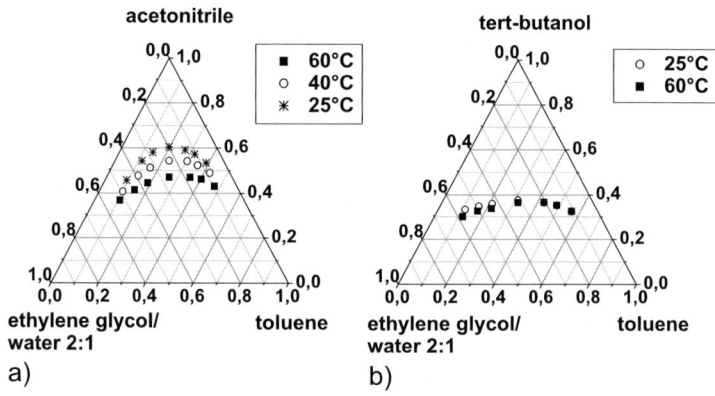

Fig. 4 Examples for solvent systems with **a** a strong and **b** a very small temperature dependency

priate solvents for the present TMS are the solvents from DMF to PEG 400 in Fig. 3.

If the non-polar solvent s2 is varied, the following tendency is observed: the less polar s2 the more mediator s3 is required until a single-phase solution is formed. Alcohols (2-octanol, 2-nonanol and 1-dodecanol), toluene, p-xylene and cyclohexane were used as s2. With DMF as the mediator a clear solution is obtained at a weight ratio s1 : 2-octanol : DMF of 1 : 3 : 0.6, while in the system s1 : toluene : DMF a ratio of 1 : 3 : 2.5 is required. With the very non-polar solvent cyclohexane the solvent mixture becomes homogeneous at a ratio s1 : cyclohexane : DMF = 1 : 3 : 11. The temperature dependency of the solvent systems s1 : s2 : DMF is almost not affected by the choice of the solvent s2.

2.1.1.2
Influence of the Reaction Components

One difficulty in the determination of an appropriate solvent system for the telomerization of butadiene with ethylene glycol is the change of polarity in the reaction mixture during the reaction.

The polar reactant ethylene glycol is consumed and as a consequence the catalyst phase becomes smaller and more polar, because of the higher polarity of water compared to ethylene glycol. At the same time a very non-polar reactant is consumed with butadiene and a large amount of semi-polar products are formed, which can also act as a mediator. As a result, more of the semi-polar solvent s3 is required at the beginning than after the reaction and the reaction mixture may no longer split up into two phases.

To investigate the influence of the reactants and the products on the solvent systems the cloud titrations were repeated with the TMS systems which show a noticeable temperature dependency. An appropriate amount of cyclohexane was added to the samples to simulate the effect of the substrate butadiene. In other tests a product mixture was investigated, which contained less ethylene glycol but an adequate amount of the semi-polar products. The required amount of s3 for the reaction mixture at the beginning of the reaction at 60 °C was compared with the one for the product mixture at 0 °C. The mediator s3 was added to the biphasic samples of ethylene glycol/water (s1) and toluene (s2) until a homogeneous solution was formed. For these investigations acetonitrile, DMSO and DMF were used as s3. With all these solvents more s3 is required when cyclohexane was added to the samples. The composition at which the solvent system s1 : toluene : DMF becomes homogeneous increases from 1 : 5 : 2.9 to 1 : 5 : 4 due to the addition of cyclohexane. In the TMS system s1 : toluene : acetonitrile the amount required rises from 1 : 5 : 4.6 to 1 : 5 : 5.5 and in the system s1 : toluene : DMSO the increase is even higher: from 1 : 3 : 3.9 to 1 : 3 : 14.4. This effect is caused by the low solubility of cyclohexane in these solvents: a second miscibility gap between s3 and cyclohexane is present in these solvent systems. The choice of cyclohexane as a substitute for butadiene was checked by repeating the cloud titrations with 1,3-pentadiene and with 1,3-butadiene itself. In all cases the same amount of the mediator s3 was required to obtain a single-phase system. For this reason and to simplify matters further experiments were performed with cyclohexane as a substitute for butadiene.

With the product mixture it was found that a much smaller amount of the mediator s3 is required to obtain a homogeneous solution than in the pure solvent mixture or the reaction mixture at the beginning of the reaction at the same temperature. For example the ratio in the TMS system s1 : toluene : DMF decreases at 60 °C from 1 : 5 : 2.9 to 1 : 5 : 2.1 and in the solvent system s1 : toluene : acetonitrile from 1 : 5 : 4.6 to 1 : 5 : 3.1. With DMSO as s3 the composition decreases from 1 : 3 : 2.4 to 1 : 3 : 3.9.

With the solvents DMF, acetonitrile and DMSO compositions can be found, which are homogeneous at reaction conditions (80 °C) and form two phases after the reaction and cooling to 0 °C, for example s1 : toluene : acetonitrile 1 : 5 : 5.5 or s1 : toluene : DMF 1 : 5 : 3.3, respectively. However, relatively large excesses of toluene and the mediator s3 are required.

2.1.1.3
Influence of the Catalyst and of the Water Content

The catalyst system Pd(acac)$_2$/TPPTS (TPPTS = trisulfonated triphenylphosphine) was used in the experiments in which the telomerization of butadiene with ethylene glycol in TMS systems was investigated. However, the catalyst precipitates from many solvent mixtures as a yellow oil or solid, as soon as a homogenous phase is obtained. For this reason the solubility of the catalyst was determined in various solvent systems. A solution of the catalyst in the mixture of ethylene glycol and water (s1) and toluene (s2) was used in a weight ratio of 1 : 3. The various mediators s3 were added until a clear solution was formed or the catalyst precipitated. Only with DMF or DMSO can a clear solution be obtained. The addition of the catalyst to the polar phase causes an increase in the amount of s3 required to achieve a homogeneous system: in the solvent system s1 : toluene : DMF the ratio increases from 1 : 5 : 4 to 1 : 5 : 4.4.

One possibility to raise the solubility of the polar catalyst in the solvent mixtures is to use a higher water content. In the TMS system s1 : toluene : DMF a larger amount of the semi-polar solvent is required to obtain a homogeneous solution, if more water is added. If the amount of water is doubled the amount of s3 increases from 1 : 5 : 4.4 to 1.35 : 5 : 6.1 and a ratio of 2 : 5 : 8.9 is needed, if the water content is four times higher. The same tendency is observed if different non-polar solvents s2 or different mediators s3 are used: the higher the water content the more of the mediator s3 is required. The temperature dependency is almost not affected when more water is added to the solvent systems.

To avoid difficulties with the low solubility of Pd/TPPTS the monosulfonated triphenylphosphine (TPPMS) was used as the ligand, which is less polar than the trisulfonated TPPTS. With the following solvents s3 a homogeneous phase is obtained: DMF, DMSO, PEG 400 and isopropyl alcohol. A smaller amount of the mediator is required if TPPMS is used than with TPPTS as the ligand. In further experiments the use of polyethylene glycol

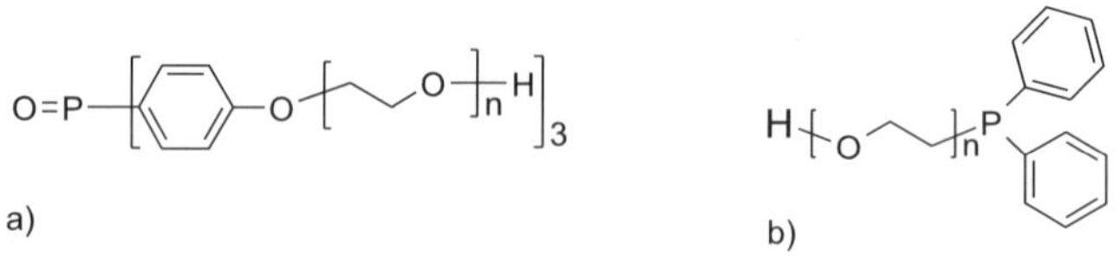

a)   b)

Fig. 5 PEG-modified ligands: **a** PETPPO and **b** MDPP

(PEG)-modified ligands [e.g. polyethylene glycol triphenylphosphine oxide (PETPPO) and methyl polyethylene glycol diphenylphosphine (MDPP)—Fig. 5] was investigated.

With both ligands a clear solution is formed with all mediators tested so far and the palladium complexes of PETPPO and MDPP are soluble in the solvent mixtures. With the exception of the alcohols and PEG 400 a smaller amount of the mediator is required to obtain a homogeneous phase as compared to the use of the sulfonated ligands.

2.1.1.4
Telomerization of Butadiene with Ethylene Glycol in TMS Systems

Catalysis experiments were performed to investigate the telomerization of butadiene with ethylene glycol in selected TMS systems (e.g. s1 : toluene : DMF 1 : 5 : 4 or s1 : 2-octanol : DMSO 1.35 : 3 : 5.2). With Pd/TPPTS as the catalyst a maximum yield of only 10% of the desired products could be achieved. With Pd/TPPMS the yield increased up to 43% in the TMS system s1 : toluene : isopropyl alcohol, but additional water had to be added to obtain a phase split after the reaction. The catalyst leaching is very high and 29% of the palladium used is lost to the product phase.

A yield of up to 30% is achieved by use of PETPPO in the TMS system s1 : toluene : acetonitrile and with MDPP as the ligand the yield reaches a level of 20%. Again, however, too much of the palladium catalyst is lost to the product phase in both cases with about 40% to 50% (Table 1).

Table 1 Telomerization of butadiene with ethylene glycol in TMS systems. Reaction conditions: 0.06 mol % Pd(acac)$_2$ based on ethylene glycol, Pd/P = 1 : 3; butadiene/ethylene glycol = 2.5 : 1, s1 = ethylene glycol : water 2 : 1, 80 °C; 4 h; 1200 rpm

TMS system	Ligand	Yield (1) [%]	Pd loss [%]
s1 : 2-octanol : DMSO 1.35 : 3 : 5.2*	TPPTS	10	not determined
s1 : toluene : isopropyl alcohol 1 : 3 : 3	TPPMS	43	29
s1 : toluene : acetonitrile 1 : 3 : 4	PETTPO	30	40
s1 : toluene : acetonitrile 1 : 3 : 4	MDPP	20	50

* 0.225 mol % Pd(acac)$_2$ based on ethylene glycol

2.1.1.5
Use of Cyclodextrins as Phase Transfer Catalysts

An alternative possibility for catalyst recycling by a phase separation is the use of phase transfer catalysts in a biphasic reaction. In the telomerization of butadiene and ethylene glycol the catalyst Pd/TPPTS is located in the aqueous phase and it is in this phase that the reaction takes place. Therefore, substances like cyclodextrins, which are well known as inverse phase transfer catalysts [11–14] might be suitable to accelerate the reaction and to improve the yield. α-Cyclodextrin and two different derivatives of β-cyclodextrin (methylated β-cyclodextrin and hydroxypropyl-β-cyclodextrin) were used in the catalysis experiments. The results obtained are shown in Table 2.

Almost no influence of the methylated β-cyclodextrin on the yield can be observed if no additional organic solvent (except for butadiene or the product) is used as the non-polar phase. With hydroxypropyl-β-cyclodextrin a decreased yield is obtained and the phase separation is more difficult. If cyclohexane or n-octane is added as the non-polar solvent the conversion and the yield increase with increasing cyclodextrin concentration up to a concentration of about 2 mol % cyclodextrin based on butadiene. Because of the high viscosity of the solution no further improvement can be achieved using higher cyclodextrin concentrations. Lower yields are obtained by use of α-cyclodextrin as compared with the methylated β-cyclodextrin.

The palladium leaching to the product phase was investigated via ICP-OES measurements. In all cases about 5% of the metal catalyst is lost. This palladium loss is in the same range as in the biphasic reaction in water with subsequent extraction with cyclohexane. Therefore, one can conclude that the use of cyclodextrins has almost no influence on the palladium leaching. For the reaction described in this work, this makes the use of cyclodextrins as PCT catalysts more attractive than the TMS systems to overcome mass transfer limitations.

Table 2 Telomerization of butadiene with ethylene glycol with addition of cyclodextrins. Reaction conditions: 0.06 mol % Pd(acac)$_2$/0.3 mol % TPPTS based on ethylene glycol; Pd/P = 1 : 5; butadiene/ethylene glycol = 2.5 : 1, s1 = ethylene glycol : water 2 : 1, 80 °C; 4 h; 1200 rpm

| | Yield (1) [%] | | |
	Cyclohexane	n-octane	No non-polar solvent
no cyclodextrine	44	52	70
0.25 mol % methylated β-cyclodextrin	52	60	72
1 mol % methylated β-cyclodextrin	65	67	70
2 mol % methylated β-cyclodextrin	57	–	–
1 mol % hydroxypropyl-β-cyclodextrin	–	–	64
0.25 mol % α-cyclodextrin	31	–	63

2.1.2
Telomerization of Butadiene with Carbon Dioxide

The telomerization of butadiene with carbon dioxide to form a δ-lactone is an interesting example for a C – C bond-forming reaction with CO_2 (Scheme 2). The product can be hydrogenated to 2-ethylheptanoic acid, which can be used in lubricants, as a stabilizer for PVC or as an intermediate for the production of solvents and softeners [7, 15–19].

In a first set of experiments we investigated the solubility of the δ-lactone in different solvents and with the distribution of the product and the catalyst between a polar and a non-polar phase. The δ-lactone itself is semi-polar in nature, thus a great variety of solvents are applicable for the extraction of the product. Two cases are possible: the use of a polar catalyst phase and a non-polar product phase or vice versa the use of a non-polar catalyst phase and a polar product phase.

As a polar solvent for the catalyst ethylene carbonate (EC), propylene carbonate (PC) and acetonitrile were used. Tricyclohexylphosphine, triphenylphosphine and the monosulfonated triphenylphosphine (TPPMS) were investigated as ligands with Pd(acac)$_2$ as the precursor. Cyclohexane, dodecane, *p*-xylene and alcohols (1-octanol, 2-octanol and 1-dodecanol) were tested as non-polar solvents for the product. To determine the distribution of the product and of the catalyst, the palladium precursor and the ligand were dissolved in the polar solvent and twice as much of the non-polar solvent was added. After the addition of δ-lactone, the amounts of the product in both phases was determined by gas chromatography. The product is not soluble in cyclohexane and dodecane, more than 99% of it can be found in the polar catalyst phase. With the alcohols 1-octanol, 2-octanol and dodecanol about 50 to 60% of the δ-lactone are located in the non-polar phase. With *p*-xylene biphasic systems can only be achieved when EC is used as the polar solvent and even in this solvent system one homogeneous phase is formed at a temperature higher than 70 °C. In a 1 : 1 mixture of EC and *p*-xylene about 50 to 60% of the product is contained in the polar phase.

The distribution of the catalyst depends on the choice of the non-polar solvent and on the ligand used. With cyclohexane and the alcohols the palladium complexes of tricyclohexylphosphine and triphenylphosphine are located in the polar phase, with *p*-xylene the complex of triphenylphosphine is dispersed in both phases, and the complex of tricyclohexylphosphine can be predominantly found in the non-polar phase. Because of the low solubility of

Scheme 2 Telomerization of butadiene with carbon dioxide

TPPMS in EC, PC and acetonitrile, no further experiments were performed with this ligand.

To investigate the use of a non-polar catalyst phase and a polar product phase EC, PC and acetonitrile were chosen as polar solvents and cyclohexane and *p*-xylene as non-polar solvents. Tricyclohexylphosphine, triphenylphosphine and bisadamantyl-*n*-butyl-phosphine were used as the ligand for Pd(acac)$_2$. If cyclohexane is used as the non-polar solvent, the palladium complexes of tricyclohexylphosphine and triphenylphosphine are situated in the polar solvent and with *p*-xylene the complex of tricyclohexylphosphine is located in the non-polar phase. In the solvent system EC/cyclohexane the palladium complex of bisadamantyl-*n*-butyl-phosphine can be found in the cyclohexane phase.

The results of this analysis of the product and catalyst distribution show that only a limited range of systems may be applicable for the telomerization of butadiene and carbon dioxide. The reaction was performed in the biphasic systems EC/2-octanol, EC/cyclohexane and EC/*p*-xylene. The yield of δ-lactone reached only 3% after a reaction time of 4 hours at 80 °C. In the solvent system EC/2-octanol triphenylphosphine was used as the ligand. With the ligand bisadamantyl-*n*-butyl-phosphine even lower yields were achieved in a single-phase reaction in EC or in the biphasic system EC/cyclohexane. The use of tricyclohexylphosphine led to a similar result, only 1% of the desired product was obtained in the solvent system EC/*p*-xylene, which forms one homogeneous phase at the reaction temperature of 80 °C. Even at a higher temperature of 100 °C and a longer reaction time of 20 hours no improvement could be observed. Therefore, we turned our interest to another telomerization-type process.

2.1.3
Carboxytelomerization

The carboxytelomerization, a variant to the usual telomerization, is the carbonylation–dimerization of butadiene, carbon monoxide and alcohols.

Catalysts described in the literature and in patents contain Pd(OAc)$_2$ and PR$_3$ (R = alkyl and aryl). Short-chained alcohols (C$_1$ to C$_4$) are typically used as nucleophiles (Scheme 3) and at reaction conditions of 120 °C, 50 bar CO and 15 hours, the desired linear (isopropyl-)nonadienoate is formed as the main product with isopropyl alcohol (IPA) as the alcohol component.

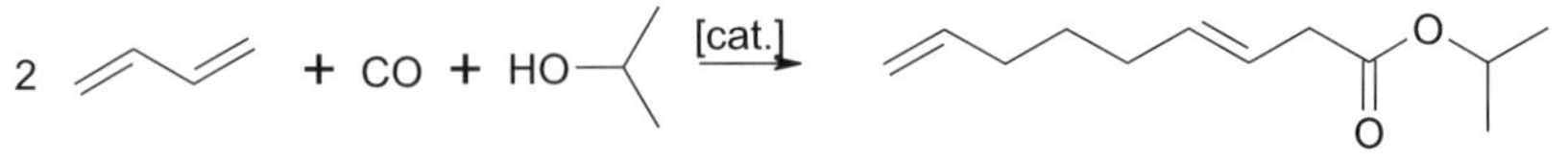

Scheme 3 Carboxytelomerization of butadiene with carbon monoxide and isopropyl alcohol

Table 3 Carboxytelomerization with commercially available ligands. Reaction conditions: 0.1 mol% Pd(OAc)$_2$/0.4 mol% ligand based on butadiene; Pd/P = 1 : 4; butadiene/isopropyl alcohol = 1 : 1, 120 °C; 15 h; 50 bar CO; 800 rpm

Entry	Ligand	Additive	Solvent	Yield [%]	Pd [%]*
1	PPh$_3$	MA	IPA	14	20
2	TPPMS	–	IPA	0	
3	TPPMS	–	IPA + 10% H$_2$O	0	
4	P(n-bu)$_3$	MA	IPA	42	62
5	P(i-bu)$_3$	–	IPA	57	80
6	P(i-pr)$_3$	–	IPA	64	
7	P(i-pr)$_3$	MA	IPA	64	98
8	P(t-bu)$_3$	–	IPA	0	
9	P(cy-hex)$_3$	MA	IPA	43	
10	P(ad)$_2$(n-bu)	–	IPA	34	94
11	P(ad)$_2$(n-bu)	MA	IPA	46	
12	bis-mes-IM	dvds	IPA	7	0.2

* Residual amount of Pd, determined by ICP

In the present study IPA was used as both nucleophile and solvent. The research was started with a screening for suitable ligands. The first ligands tested were commercially available (Table 3).

The results show that basic alkylphosphines are especially suitable. The ligand should be sterically demanding, but tri-t-butylphosphine (entry 8) is obviously too bulky. An ideal ligand seems to be tri-i-propylphosphine (entries 6 and 7). Unfortunately, the water-soluble ligands TPPMS (entries 2 and 3) and TPPTS do not work in this reaction, also the carbene ligand bis(mesitylene)imidazolium chloride (entry 12) has only a low activity. The influence of additives like maleic anhydride (MA) and 1,3-divinyltetramethyldisiloxane (dvds) is negligible.

From our cooperation partners, Profs. Gladysz and Dinjus, we received ligands with perfluorinated chains ("ponytails"), which show a thermomorphic solubility in organic solvents (P(et-Rf$_8$)$_2$(m-me-bz)) or may be extracted with fluorous solvents (P(et-Rf$_8$)$_3$). P(et-Rf$_6$)(i-pr)$_2$ with only one perfluorinated

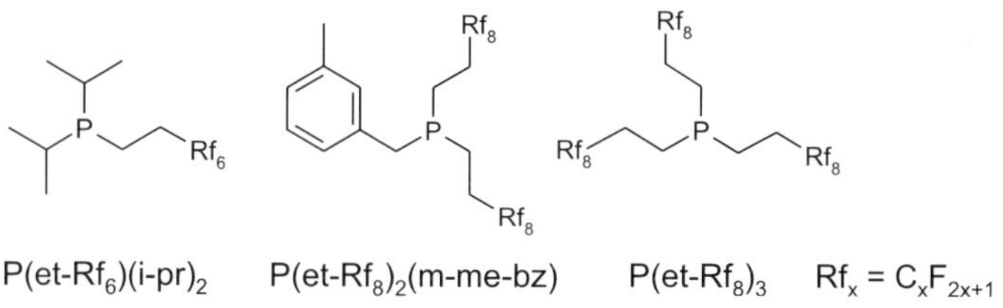

P(et-Rf$_6$)(i-pr)$_2$ P(et-Rf$_8$)$_2$(m-me-bz)   P(et-Rf$_8$)$_3$ Rf$_x$ = C$_x$F$_{2x+1}$

Scheme 4 Fluorous ligands

chain is soluble in the usual organic solvents like IPA. The structures of the ligands are shown in Scheme 4.

The influence of the degree of fluorination on the separability of the ligands was investigated. The catalyst, $Pd(OAc)_2$ with ligands, was solved in a mixture of IPA/PF-hexane. After the phase separation the Pd-leaching was measured by ICP (Table 4).

These results show that the ligand ($P(et-Rf_8)_3$) has a very low leaching in perfluorohexane. Several fluorous solvents were tested for their critical solution temperature (CST) with the substrates (Table 5).

Only perfluoromethylcyclohexane (PF-MCH, entries 17, 20, 23) is miscible both with a mixture of product and isopropyl alcohol. So all further experiments are done with this solvent.

To investigate the influence of butadiene on the CST special thick-walled glass tubes were used (entries 25–27). The behaviour of butadiene is interesting: butadiene is, unlike most of the hydrocarbons, miscible with PF-MCH (entry 25) even at room temperature. This behaviour decreases the CST of a reaction mixture in the beginning (entry 26) drastically.

Table 4 Pd-leaching in perfluorohexane with fluorous ligands

Entry	Ligand	Pd in PF-hexane-phase
13	$P(et-Rf_6)(i-pr)_2$	0%
14	$P(et-Rf_8)_2(m-me-bz)$	78%
15	$P(et-Rf_8)_3$	99%

Table 5 Critical solution temperatures of reactands with fluorous solvents

Entry	System	Vol.-ratio	CST [°C]
16	IPA : PF-octane	50 : 50	111
17	IPA : PF-MCH	50 : 50	91
18	IPA : PF-me-decaline	50 : 50	115
19	prod. : PF-octane	50 : 50	$\gg 120$
20	prod. : PF-MCH	50 : 50	> 120
21	prod. : PF-me-decaline	50 : 50	$\gg 120$
22	IPA : prod. : PF-octane	40 : 20 : 40	> 120
23	IPA : prod. : PF-MCH	40 : 20 : 40	106
24	IPA : prod. : PF-me-decaline	40 : 20 : 40	> 120
25	PF-MCH : butadiene	50 : 50	RT
26	IPA : PF-MCH : butadiene	40 : 20 : 40	32
27	IPA : prod. : PF-MCH : butadiene*	23 : 30 : 20 : 27	53

* at 25 bar CO

Table 6 Experimental results with fluorous ligands

Entry	Ligand	Solvent	Yield [%]	Pd [%]*
28	$P(et\text{-}Rf_6)(i\text{-}pr)_2$	IPA	12	55
29	$P(et\text{-}Rf_8)_2(m\text{-}me\text{-}bz)$	IPA	1.3	0
30	$P(et\text{-}Rf_8)_3$	IPA + PF-hexane	0	0
31	$P(et\text{-}Rf_8)_3$	IPA + PF-MCH	2.4	15

* Residual amount of Pd, determined by ICP

The reaction product has a converse effect (entries 22–24, 27), it increases the CST. Entry 27 shows a reaction mixture with a conversion of about 30%. This solution still becomes single-phase at moderate temperatures. During the reaction the amount of product increases while the amount of butadiene decreases. The consequence is a rising CST. To realize a single-phase during the reaction in spite of high yields it is possible to exchange the organic solvent IPA with butadiene.

The ligands of Table 4 were investigated in the carboxytelomerization reaction using the solvents IPA and IPA/PF-hexane (Table 6). The ligand $(P(et-Rf_8)_3)$, which showed the best separability, yielded the worst performance (entry 30). We also tested the reaction in the best solvent system of Table 5, the system IPA/PF-hexane. However, as shown in entry 31, the stability of Pd was improved but was still insufficient. The solvent system works quite well, but the main problem remains the activity of the ligand.

2.2
Synthesis of 4-Nitrodiphenylamine

This transformation avoids problems with the change of polarity during the reaction, which occurred in the telomerization, because two aromatic compounds react with each other to form a new aromatic product. The synthesis of 4-nitrodiphenylamine via a Pd-catalyzed Buchwald–Hartwig-type amination from 4-chloronitrobenzene and aniline was chosen as the next test reaction in a cooperation with Lanxess as industrial partner of the network (Scheme 5).

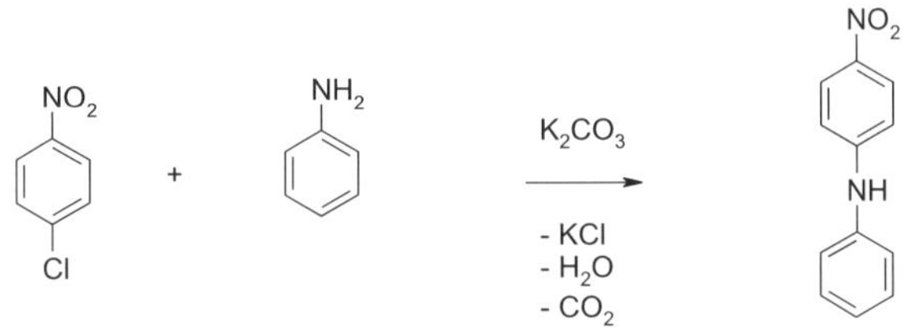

Scheme 5 Synthesis of 4-nitrodiphenylamine

Water was used as the catalyst phase for the palladium complex of TPPTS and toluene or an excess of the substrate aniline served as the non-polar product phase. To determine an appropriate solvent system cloud titrations were performed at 90 °C, 60 °C, 40 °C and 25 °C. A solution of 4-chloro-nitrobenzene in aniline and water were mixed in a weight ratio of 1 : 1 and semi-polar solvents were added as a mediator until a homogeneous solution was formed. As the mediator the following solvents were applied: methanol, ethanol, isopropyl alcohol, n-butanol, DMF, DMSO, ethylene glycol, N-methylpyrrolidone (NMP), 1.4-dioxane and acetonitrile. The cloud titrations were repeated whereby the substrate 4-chloro-nitrobenzene was replaced with the product 4-nitrodiphenylamine. In all cases more of the semi-polar mediator is required for the product mixture at 25 °C than for the reaction mixture at 60 °C to obtain a clear solution.

However, if the catalyst and the base required for the reaction were added, a large amount of the mediator is required to obtain one single phase and the solvent systems are no longer temperature dependent. Hence, in all cases a homogeneous solution is obtained after the reaction and the catalyst can not easily be recycled by a simple phase separation. This observation emphasizes again the importance of investigating the phase behaviour of the true reactive systems whenever possible.

With ethanol and DMSO as mediators catalysis experiments were performed. By use of DMSO, about 70% of the product can be obtained, if the reaction takes place in one single phase; in a two-phase system the yield decreases to about 30%. With ethanol almost no product can be detected, because a biphasic system was formed with this solvent under all conditions. In all cases the inorganic components K_2CO_3 and KCl were insoluble in the reaction mixture.

Another possibility to recover the catalyst is an extraction of the palladium and the ligand into the polar water phase followed by re-extraction with aniline into the substrate phase. Palladium can be extracted from the basic reaction mixture with water and in the product phase only a very small amount of the metal can be detected. Little catalyst decomposition is observed and a part of it remains in the solid formed from K_2CO_3 and KCl. The distribution of the catalyst depends on the water content: if more water is added more of the palladium is extracted to the water phase. After the addition of HCl the palladium can be extracted again with aniline and no metal can be found in the aqueous phase after the extraction.

Attempts to tune the distribution of the catalyst further by replacing the sodium cation of TPPTS with an anilinium cation were unsuccessful. Only a yield of about 15% could be achieved and the extraction of the palladium to the water phase was incomplete.

2.3
Isomerizing Hydroformylation of *trans*-4-Octene to *n*-Nonanal

2.3.1
The Rhodium/BIPHEPHOS Catalytic System

With an annual production of up to 9.3 million tons in 1998, hydroformylation is the most important homogeneously catalyzed reaction [20, 21]. The
reaction is performed almost exclusively by the use of cobalt or rhodium
catalysts. The advantages of rhodium catalysts are milder reaction conditions and better *n/iso* ratios in product distribution. The toxicity of rhodium
compounds as well as the high rhodium price [22] (between 20 and $75 \, \text{€} \, \text{g}^{-1}$
during the last five years) demand an efficient catalyst recycling.

Recently, we reported that the rhodium/BIPHEPHOS-catalyzed hydroformylation of *trans*-4-octene (Scheme 6) provides an interesting approach
for the synthesis of *n*-nonanal [23]. In this context *trans*-4-octene can also
be seen as a model substance for hydroformylation of internally unsaturated
fatty acid esters. This could open up access to the use of renewable resources
for the synthesis of valuable *n*-aldehydes.

Using mild reaction conditions (10 bar, 125 °C), a high conversion of *trans*-
4-octene and a high selectivity to *n*-nonanal can be obtained with toluene as
the solvent. Cyclic carbonates like propylene carbonate (PC) are also suitable
solvents for the isomerizing hydroformylation of *trans*-4-octene. Furthermore, the selectivity to *n*-nonanal is increased up to 95% when PC is used in
a single phase. The product *n*-nonanal can be extracted with *n*-dodecane or
with a mixture of dodecane isomers.

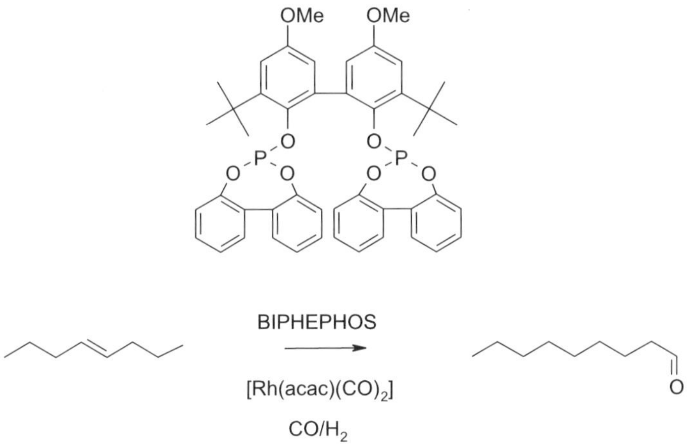

Scheme 6 Isomerizing hydroformylation of *trans*-4-octene to *n*-nonanal

2.3.2
Two-Phase Catalysis with In-Situ Extraction

Propylene carbonate is a good solvent of the rhodium precursor [Rh(acac) $(CO)_2$] and the phosphite ligand BIPHEPHOS and can thus be used as the catalyst phase in the investigation of the isomerizing hydroformylation of *trans*-4-octene to *n*-nonanal in a biphasic system [24]. As already mentioned, the reaction products can be extracted with the hydrocarbon dodecane. Instead of an additional extraction after the catalytic reaction, we carried out in-situ extraction experiments, where the products are separated from the catalytic propylene carbonate phase while the reaction is still in progress. Conversion of 96% and selectivity of 72% was achieved under comparably mild conditions ($p(CO/H_2) = 10$ bar, $T = 125\,°C$, 4 h, substrate/Rh = 200 : 1).

Upon variation of the stirring velocity between 500 and 1500 rpm the conversion of the olefin remained at the same high level and the selectivity to the linear aldehyde also remained constant. Obviously there is no mass transfer limitation in this two-phase reaction system. In comparison to the single-phase reaction in propylene carbonate as the only solvent [23], the selectivity decreases from 95% to 70%, which can be explained by the high concentration of the non-electron-donating solvent dodecane in the propylene carbonate phase. The presence of the dodecane leads to a decrease of the isomerization velocity, which results in a lower linearity of the formed aldehydes.

2.3.3
Influence of Methylated β-Cyclodextrin

Cyclodextrins are often used as inverse phase transfer catalysts [11–14]. They are able to intercalate hydrophobic substances and to transport them into a polar phase like water, where the reaction takes place. To study the influence of cyclodextrins on the isomerizing hydroformylation of *trans*-4-octene in the biphasic solvent system propylene carbonate/dodecane, the concentration of methylated β-cyclodextrin was varied from 0.2 up to 2.0 mol.-% relative to the substrate *trans*-4-octene [24]. The results are given in Table 7.

With increasing concentration of methylated β-cyclodextrin the selectivity to *n*-nonanal increases from 64% to 72%, while the conversion of the olefin is constantly as high as 97%. Obviously the addition of the methylated β-cyclodextrin has only a moderate influence on the isomerizing hydroformylation of *trans*-4-octene to *n*-nonanal. The addition of only 0.2 mol.-% of methylated β-cyclodextrin lowers the isomerization rate which results in the formation of slightly more branched aldehydes. In pharmacy β-cyclodextrins are established as solvation mediators between polar and less polar solvents. This is one possible explanation for the rise in selectivity to *n*-nonanal with an increasing β-cyclodextrin concentration. At higher con-

Table 7 Influence of the addition of methylated β-cyclodextrin in the two-phase system propylene carbonate/dodecane. Reaction conditions: 0.1 mmol [Rh(acac)(CO)$_2$], 0.5 mmol BIPHEPHOS, 19.4 mmol *trans*-4-octene, 20 ml propylene carbonate, 20 ml dodecane, p(CO/H$_2$ = 1/1) = 10 bar, T = 125 °C, t = 4 h, stirring velocity 500 rpm

Concentration β-cyclodextrin	Conversion (*trans*-4-octene) [%]	Selectivity (*n*-nonanal) [%]
0.2 mol.-%	97	64
1.0 mol.-%	97	68
2.0 mol.-%	96	72

centration the former two-phase reaction system changes to a single-phase reaction system which leads to a higher linearity of the aldehydes.

2.3.4
Isomerizing Hydroformylation in TMS Systems

To elucidate the use of TMS systems for the isomerizing hydroformylation, PC was chosen as the solvent for the rhodium catalyst, because the best selectivity to *n*-nonanal of 95% with a conversion on *trans*-4-octene of also 95% was achieved in this solvent under single-phase conditions. Dodecane was used as a non-polar solvent for the extraction of the product and *p*-xylene served as the mediator between the catalyst and the product phase [24]. Appropriate operation points for the reaction within this solvent system were determined by cloud titrations.

The TMS system PC/dodecane/*p*-xylene shows the phase behaviour depicted in Fig. 6 representing a system with a closed miscibility gap, which shows a strong temperature dependence. Possible solvent compositions are defined by the area between the two binodal curves at the temperatures of 25 °C and 80 °C.

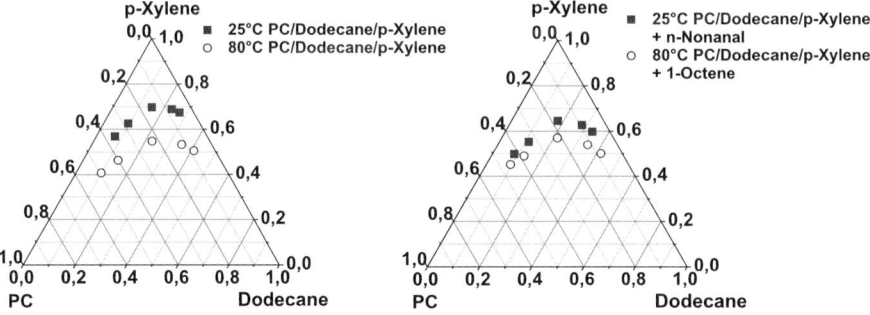

Fig. 6 Solvent system propylene carbonate (PC)/dodecane/*p*-xylene

To investigate the influence of the starting olefin and the generated aldehydes in the temperature-controlled system a constant amount of *trans*-4-
octene was added to the solvent system (0.15 g *trans*-4-octene/g of the two-
phase system PC/dodecane) and the measurements were repeated at 80 °C,
the starting point of the reaction. To ensure phase separation after the reaction, we added *n*-nonanal to the solvent mixture (0.17 g *n*-nonanal/g of
the two-phase system PC/dodecane) and measured it again at 25 °C. This
more realistic TMS-system is presented on the right side of Fig. 6. As can
be seen the position of the binodal curve is not affected by the addition of
trans-4-octene. On the other side, the addition of *n*-nonanal has a significant
influence on the solvent system. The two binodal curves move closer to each
other which results in a diminished working area.

The isomerizing hydroformylation of *trans*-4-octene has been executed
in PC/dodecane/*p*-xylene with varying compositions of the three solvents.
The phase diagram with the corresponding working points is presented
in Fig. 7.

The conversion of *trans*-4-octene is very high in this TMS-system and
reaches a level of 99%. The selectivity to the desired linear aldehyde amounts

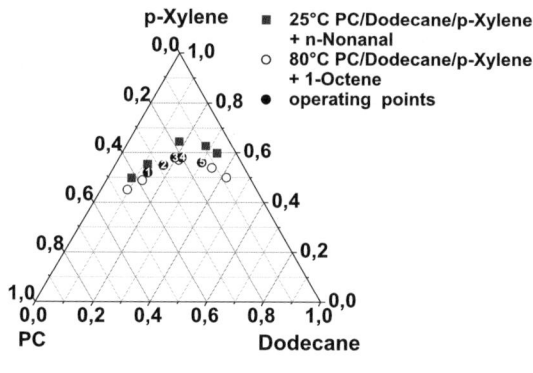

PC/Dodecane/*p*-Xylene [wt.-%]	Selectivity (*n*-Nonanal) [%]
Point 1: 35/13/52	90
Point 2: 28/17/55	86
Point 3: 22,5/19.5/58	90
Point 4: 20/22/58	79
Point 5: 14/30/56	82

Fig. 7 Isomerizing hydroformylation of *trans*-4-octene in the TMS system PC/dodecane/
p-xylene

to about 90% and is higher compared to the two-phase catalysis with in-situ extraction or with addition of methylated β-cyclodextrin. The selectivity to *n*-nonanal is dependent on the concentration of PC: the higher the concentration of propylene carbonate, the higher the selectivity to *n*-nonanal. This significant influence of the PC on the selectivity may be associated with the high polarity of the carbonate group.

However, the TMS-system PC/dodecane/*p*-xylene has still some severe limitations. Via ICP-investigations a strong rhodium leaching of 47% of the rhodium catalyst was detected. Furthermore, we observed a correlation between the amount of the mediator *p*-xylene and the amount of leaching. The more *p*-xylene used, the more rhodium is transferred into the non-polar dodecane phase. Therefore, catalyst recycling in these systems is impossible at the moment.

For this reason pyrrolidones were tested as semi-polar solvents in further experiments [25] and ethylene carbonate (EC) and butylene carbonate (BC) were examined as the polar catalyst phase instead of PC.

N-octylpyrrolidone (NOP) unites polar and non-polar properties and was used as a mediator between the cyclic carbonates and *n*-dodecane as the extraction agent. Figure 8 shows the mixing behaviour of some TMS systems for the different cyclic carbonates. Obviously the miscibility gap increases with decreasing length of the carbonate's carbon chain.

Both the cyclic carbonates and NOP are able to dissolve the catalyst. It was observed that at separation temperature NOP can be found in the product phase as well as in the catalyst phase. This causes again undesired catalyst leaching (Table 8) albeit at levels that could be reduced already down to 1% with BC as the polar phase. Thus, it was necessary to identify a mediator with a similar property like NOP that exclusively can be found in the catalyst phase.

As a possibility to classify the mixing behaviour of solvents, Hildebrand and Scott [26, 27] developed the theory of the solubility parameter (δ). It is

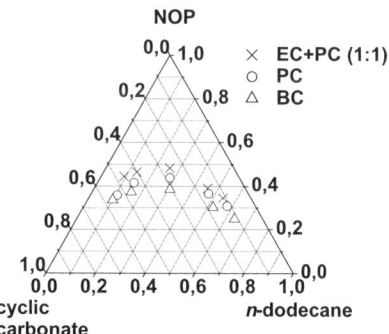

Fig. 8 Mixing behaviour of some cyclic carbonates (EC, PC, BC) with *n*-dodecane and *N*-octylpyrrolidone (NOP) at 25 °C

Table 8 Catalyst leaching dependent of cyclic carbonate

Cyclic carbonate [wt.%]	n-dodecane [wt.%]	NOP [wt.%]	Rh leaching [%]	P leaching [%]
EC 50	12	38	90	60
EC 25 + PC 25	23	27	16	11
PC 50	23	27	2	2
BC 50	31	19	1	2

defined as the square root of the cohesive energy density:

$$\delta = \sqrt{\frac{E}{V}}, \tag{1}$$

where E is the energy of vaporization of a pure solvent (reduced by the potential energy of an ideal gas) with the molar volume V. The theory postulates that two solvents are more likely to be miscible if their δ values are similar.

Hansen [28, 29] expanded this theory by dividing the cohesive forces of liquids into three components—dispersive (d), polar (p) and hydrogen bonding (h) forces—and defined the three component solubility parameter δ_0 as:

$$\delta_0 = \sqrt{\delta_d + \delta_p + \delta_h}. \tag{2}$$

In the context of the TMS investigated here, a mediator was needed with a Hansen solubility parameter (HSP) closer to that of the catalyst solvent than to that of NOP. A detailed database of HSPs can be found in the literature [30–32].

As stated previously the main criteria in solvent selection for extraction purposes is the polarity. To prevent miscibility of the mediator with the product phase at separation temperature the mediator's polarity has to be preferably in the range of the cyclic carbonate. Table 9 shows that with decreasing chain length of the N substituent of a pyrrolidone cycle, the solubil-

Table 9 Hansen solubility parameters of some solvents

Solvent	δ_0 [MPa^{0,5}]	δ_d [MPa^{0,5}]	δ_p [MPa^{0,5}]	δ_h [MPa^{0,5}]
n-dodecane	16.0	16.0	0.0	0.0
ethylene carbonate	29.6	19.4	21.7	5.1
propylene carbonate	27.3	20.0	18.0	4.1
N-benzylpyrrolidone	20.0	18.2	6.1	5.6
N-n-butylpyrrolidone	20.9	17.5	9.9	5.8
N-methylpyrrolidone	22.9	18.0	12.3	7.2

ity parameter increases. In comparison to NOP, N-methylpyrrolidone (NMP) is a common solvent in the chemical industry and therefore much cheaper. Together with the higher HSP, this makes NMP an interesting substitute for NOP. Figure 9 shows the substitution effects of NMP.

In general, more NMP than NOP is needed to close the miscibility gap between the cyclic carbonate and the extraction agent. In addition, a higher temperature dependency of the mixing behaviour can be recognized upon use of NMP: at room temperature the NMP-TMS systems pass from closed to open systems. This effect results in an almost complete immiscibility of the mediator and the cyclic carbonate with the extraction agent at separation temperature. In this way the catalyst leaching into the product phase can be suppressed.

The determination of suitable operating points demands the consideration of the influence of the reaction's educts and products. We ascertained that in this case the miscibility gap is increased by *trans*-4-octene at reaction temperature and decreased by nonanal at separating temperature which causes a diminution of the operating area. This effect is illustrated in Fig. 10.

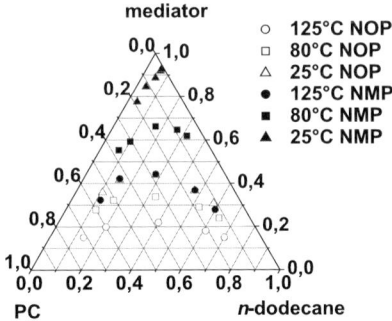

Fig. 9 Comparison of *N*-octylpyrrolidone (NOP) with *N*-methylpyrrolidone (NMP) as mediator for propylene carbonate (PC) and *n*-dodecane

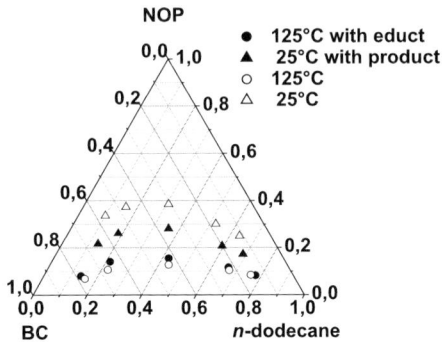

Fig. 10 Effect of *trans*-4-octene and *n*-nonanal on the mixing behaviour of TMS; addition of 15 wt % *trans*-4-octene at reaction temperature (125 °C) and 17 wt % nonanal at separating temperature (25 °C)

The hydroformylation was investigated at various compositions of the TMS system propylene carbonate/*n*-dodecane/*N*-octyl-2-pyrrolidone (16/63/21, 24/50/26, 36/36/28, 50/23/27, 63/13/24 wt. %). With increasing mass fraction of propylene carbonate the conversion of *trans*-4-octene can be increased from 92 to 98% while the selectivity to *n*-nonanal decreases from 81 to 72%. In the same way the rhodium loss is reduced from 21 to 1% and the loss of phosphorous from 15 to 7% as compared to similar conditions with NOP.

The variation of the cyclic carbonate revealed that with increasing polarity (decrease in carbon chain length) the conversion regarding *trans*-4-octene decreases from 100 to 94% and the selectivity to *n*-nonanal increases from 76 to 86%. This is accompanied with rising catalyst leaching which reflects again the increasing difference in HSP between the cyclic carbonate and the mediator (Table 10).

Comparable experiments with *N*-methyl-2-pyrrolidone as the mediator produced similar results, but now with an even lower catalyst leaching between 0.06 and 0.26% for rhodium and 0.56 and 1.40% for phosphorous

Table 10 Hydroformylation of *trans*-4-octene in cyclic carbonate/*N*-octyl-2-pyrrolidone/ extraction agent, $T = 125\,^{\circ}C$, p(syngas, $1:1$) = 10 bar, t(reaction) = 4 h, m(solvents) = 30 g, n([Rh(acac)(CO)$_2$]) = 10^{-4} mol, n(BIPHEPHOS) = 5×10^{-4} mol

Cyclic carbonate [wt.%]	Extraction agent [wt.%]	NOP [wt.%]	Conversion of *trans*-4-octene [%]	Selectivity to *n*-nonanal [%]	Rh leaching [%]	P leaching [%]
EC 50	*n*-Do 12	38	94	86	89.53	60.34
EC 25 + PC 25	*n*-Do 23	27	97	80	12.23	6.63
PC 50	*n*-Do 23	27	98	79	2.48	2.07
BC 50	*n*-Do 31	19	100	76	1.19	2.36

Table 11 Hydroformylation of *trans*-4-octene in cyclic carbonate/*N*-methyl-2-pyrrolidone/ extraction agent, $T = 125\,^{\circ}C$, p(syngas, $1:1$) = 1 MPa, t(reaction) = 4 h, m(solvents) = 30 g, n([Rh(acac)(CO)$_2$]) = 10^{-4} mol, n(BIPHEPHOS) = 5×10^{-4} mol

Cyclic carbonate [wt.%]	Extraction agent [wt.%]	NMP [wt.%]	Conversion of *trans*-4-octene [%]	Selectivity to *n*-nonanal [%]	Rh leaching [%]	P leaching [%]
EC 18 + PC 18	*n*-Do 10	54	99	80	0.06	0.56
PC 50	*iso*-Do 10	40	99	81	0.04	0.51
BC 50	*n*-Do 23	27	100	76	0.26	1.40

(Table 11). The system performance was very stable, regardless of the mediator or extracting agent. These results demonstrate that the TMS approach can indeed lead to highly active and selective processes with a remarkably efficient separation.

2.4
TMS Systems for Hydroaminomethylation

Hydroaminomethylation is a simple, efficient and atom-economic method to synthesize various amines. This one-pot reaction consists of three consecutive steps: in the first step a hydroformylation of an olefin is performed followed by the reaction of the resulting aldehyde with a primary or secondary amine to give the corresponding enamine or imine. Lastly, this intermediate is hydrogenated to the desired secondary or tertiary amine (Fig. 11) [33–39]. In most cases rhodium salts or complexes are used as the homogeneous catalyst in the hydroaminomethylation.

TMS systems, which were used in the isomerizing hydroformylation of *trans*-4-octene, should be applicable to hydroaminomethylation as well because the hydroformylation is the first step of the reaction. For this reason similar TMS systems were applied in a first series of investigations [40]. Propylene carbonate (PC) was chosen as the polar solvent s1 for the catalyst and alkanes (an isomeric mixture of dodecane or *n*-hexane) were used as non-polar component s2. 1.4-Dioxane, different pyrrolidones [*N*-methylpyrrolidone (NMP), *N*-ethylpyrrolidone (NEP), *N*-cyclohexylpyrrolidone (NCP) *N*-benzylpyrrolidone (NBP) and *N*-octylpyrrolidone (NOP)] or esters of lactic acid (ethyllactate and butyllactate) served as mediator s3. As a test reaction the hydroaminomethylation of 1-octene with morpholine was investigated (Scheme 7).

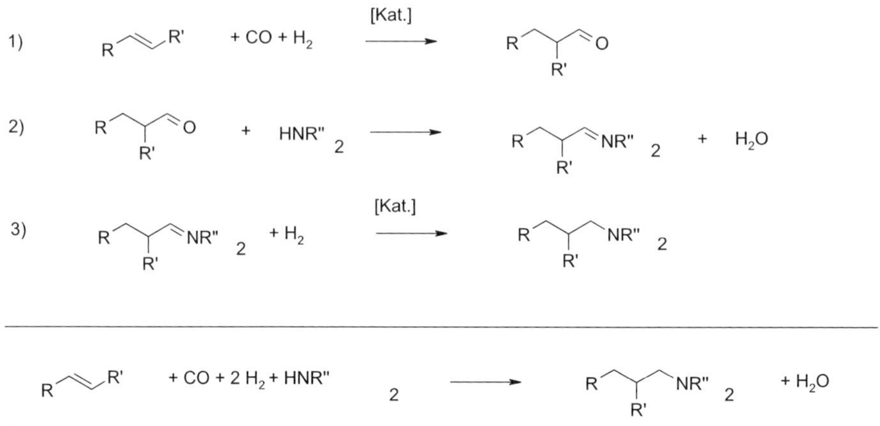

Fig. 11 Reaction steps of the hydroaminomethylation

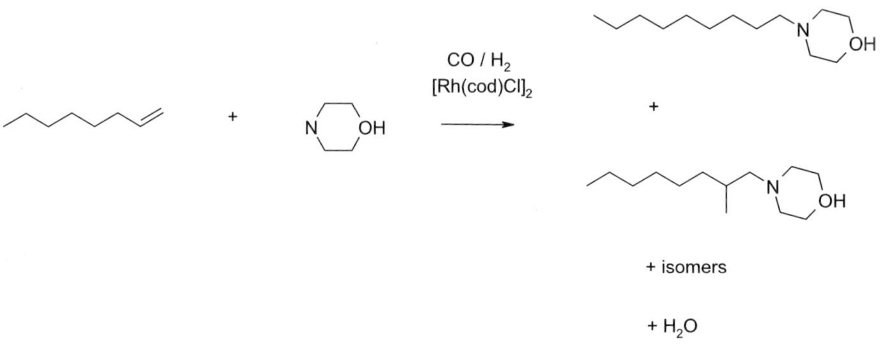

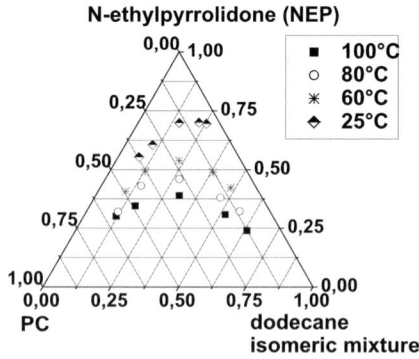

Scheme 7 Hydroaminomethylation of 1-octene with morpholine

To determine appropriate TMS systems cloud titrations were performed at different temperatures from 25 °C to 100 °C. To a mixture of PC and s2 the mediator s3 was added dropwise until a homogeneous solution was formed. The phase behaviour of the TMS system PC/dodecane/NEP is shown in Fig. 12.

The system is exceptionally temperature dependent and the miscibility gap decreases with rising temperature. With the other pyrrolidones the required amount of s3 decreases with increasing length of the alkyl chain at the pyrrolidone and the decreasing polarity of the mediator involved.

With N-benzylpyrrolidone (NBP) great amounts (> 80% of the mixture) of the solvent are required. Therefore, no further experiments were performed with NBP. The temperature dependency decreases in the same order: NMP > NEP > NCP > NOP (Fig. 13).

The TMS systems PC/dodecane/NMP and PC/dodecane/NEP pass from a system with a closed miscibility gap to one with an open miscibility gap at room temperature. This immiscibility of the mediator and the polar phase with the product phase can be used for a convenient recovery of the catalyst.

Fig. 12 Thermomorphic solvent system PC/dodecane/NEP

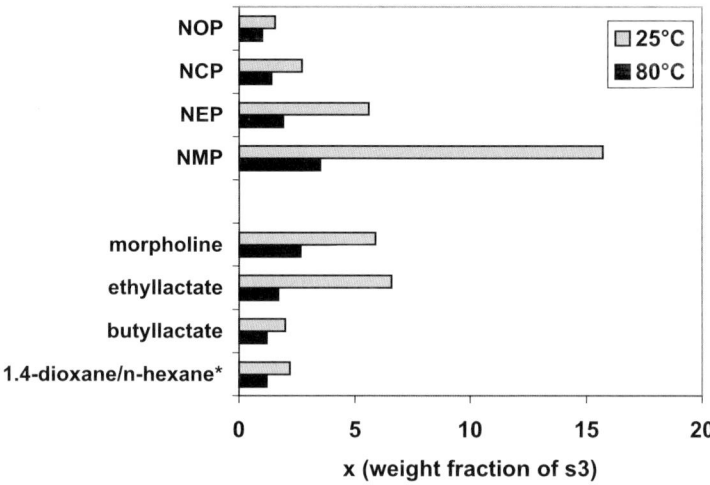

Fig. 13 Required weight ratio PC/dodecane/s3 = 1 : 1 : x for different mediators s3* PC/
n-hexane/1.4-dioxane = 1 : 0.55 : x

The influence of the substrates 1-octene and morpholine was measured by repeating the cloud titrations with addition of these substances to determine the appropriate composition of the solvent mixture for the reaction. While the addition of 1-octene has almost no effect on the TMS systems the addition of morpholine leads to a decrease of the miscibility gap (Fig. 14). As morpholine itself is semi-polar and can be used as a mediator a smaller amount of s3 is needed to obtain a homogeneous system.

The application of morpholine as s3 was studied in a further test run. In this case a ratio of PC : dodecane : morpholine of 1 : 1 : 2.6 by weight is required to get a clear solution at 80 °C (Fig. 13). The TMS system PC/

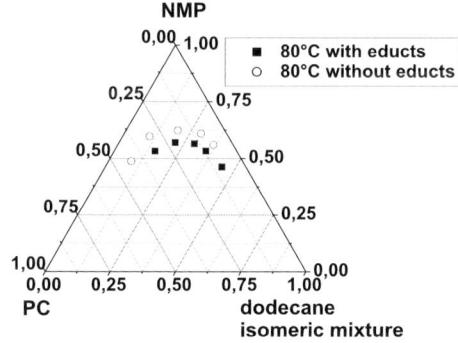

Fig. 14 Influence of the substrates 1-octene and morpholine on the TMS system PC/
dodecane/NMP

dodecane/morpholine shows a strong temperature dependence. With 1.4-dioxane and the lactates smaller amounts of s3 are necessary. For example a weight ratio of $1:1:1.7$ is efficient in the TMS system PC/dodecane/butyllactate. With the lactates as mediator the same tendency is observed as with the pyrrolidones: with increasing chain length and polarity both the amount needed to get a homogeneous phase and the temperature dependency of the solvent systems decrease. Similar to the systems with NMP or NEP as s3 the TMS system PC/dodecane/ethyllactate changes from a system with a closed miscibility gap to a system with an open miscibility gap at room temperature. In all cases a smaller amount of s3 is required, when the educt morpholine is added, because of its semi-polar nature.

The hydroaminomethylation of 1-octene and morpholine was carried out in PC as a single phase and in the TMS systems described above. PC and s2 were used in a weight ratio of $1:1$ and an appropriate amount of s3 was added to get a homogeneous phase at reaction temperature. As the catalyst $[Rh(cod)Cl]_2$ was used without any further ligand. The results obtained are shown in Table 12.

A high conversion of 1-octene up to 98% after two hours at $125\,°C$ is observed in all cases. The selectivity to the amines is also very high, it reaches a level of 96% in the TMS system PC/n-hexane/dioxane. Without using a ligand none of the regioisomers is clearly preferred under the given reaction

Table 12 Hydroaminomethylation of 1-octene and morpholine in different thermomorphic solvent systems. Reaction conditions: $0.1\,mol\%$ $[Rh(cod)Cl]_2$ based on 1-octene, morpholine/1-octene $1.5:1.8\,bar\,CO$, $39\,bar\,H_2$, $125\,°C$, $2\,h$, stirrer velocity $1000\,rpm$

Solvent system	Conversion 1-octene [%]	Amine selectivity [%]	$n:iso$	TON	TOF [h^{-1}]	Rh loss [%]
PC	97	71	2.0 : 1	970	485	–
PC/n-hexane/1.4-dioxane 1 : 0.55 : 1.3	97	96	1.4 : 1	970	485	0.7
PC/dodecane/NMP 1 : 1 : 2.65	95	68	3.0 : 1	950	475	0.4
PC/dodecane/NOP 1 : 1 : 0.85	92	92	1.6 : 1	920	460	1.5
PC/dodecane/NEP 1 : 1 : 1.2	97	78	2.1 : 1	970	485	1.1
PC/dodecane/morpholine 1 : 1 : 2.8	98	69	1.8 : 1	980	490	0.4
PC/dodecane/ethyllactate 1 : 1 : 1.5	79	50	2.1 : 1	790	395	0.9
PC/dodecane/butyllactate 1 : 1 : 1	82	60	1.7 : 1	820	410	0.8

conditions, so that only low *n/iso* ratios can be observed. The product is located in both phases with the distribution depending on the mediator. With more polar solvents s3, more of the mediator and of the product can be found in the catalyst phase. Only very small amounts of the possible by-products nonanal, nonanol or *n*-octane can be detected.

However, morpholine-4-carboxylic acid 2-hydroxy-1-methyl-ethyl ester is formed by the reaction of PC and the substrate morpholine in an undesired side reaction. By use of 1.4-dioxane or the pyrrolidones as mediator s3 about 30 to 45% of the morpholine is consumed by this side reaction. The by-product is contained in the PC phase and can not be extracted to the non-polar product phase. The selectivity to the desired amines is lowered, because of the consumption of the morpholine. Thus, PC has to be substituted by another polar solvent (e.g. water, methanol or ethylene glycol) in future experiments. The lactates react with the morpholine, too resulting in the corresponding amide. Overall, the hydroaminomethylation in the TMS systems PC/dodecane/lactate results in a conversion of 1-octene of about 80%, but in selectivities to the amines of only 50 to 60%.

The rhodium loss to the product phase was investigated by ICP-OES spectrometry. In all cases very low catalyst leaching can be observed and less than 1.5% of the rhodium is extracted (Table 12). The catalyst loss can be correlated to the polarity and the solubility of the mediator s3 in the product phase s2. The less polar s3, the more s3 is dissolved in s2 and the more catalyst is lost.

Table 13 Bergs classification of solvents

Class 1	Molecules, which may create a three-dimensional net of strong H-bondings	H_2O, polyols, aminoalcohols, oxy-acids, polyphenols, hydroxylamine etc.
Class 2	Molecules with active H-atoms and electronegative atoms with pairs of free electrons (O, N, F)	Alcohols, acids, phenols, prim. and sek. amines, oximes, nitro compounds and nitriles with α-H-atoms like hydrazine, NH_3, HF, HCN, etc.
Class 3	Molecules without active H-atoms but with electronegative atoms	Ethers, ketones, aldehydes, esters, tert. amines, pyridine, nitro compounds and nitriles without α-H-atoms
Class 4	Molecules with active H-atoms but without electronegative atoms	Molecules with 2 or 3 Cl-atoms at the C-atom with an active H-atom or with one Cl at this C and Cl-atoms at an adjacent C e.g. $CHCl_3$, CH_2Cl_2, $CH_2Cl - CH_2Cl$
Class 5	Molecules which are unable to form H-bondings or molecules with large unpolar substituents with a higher influence than the polar substituents	Hydrocarbons, CS_2, CCl_4, sulfides, mercaptanes, nonmetallic elements etc.

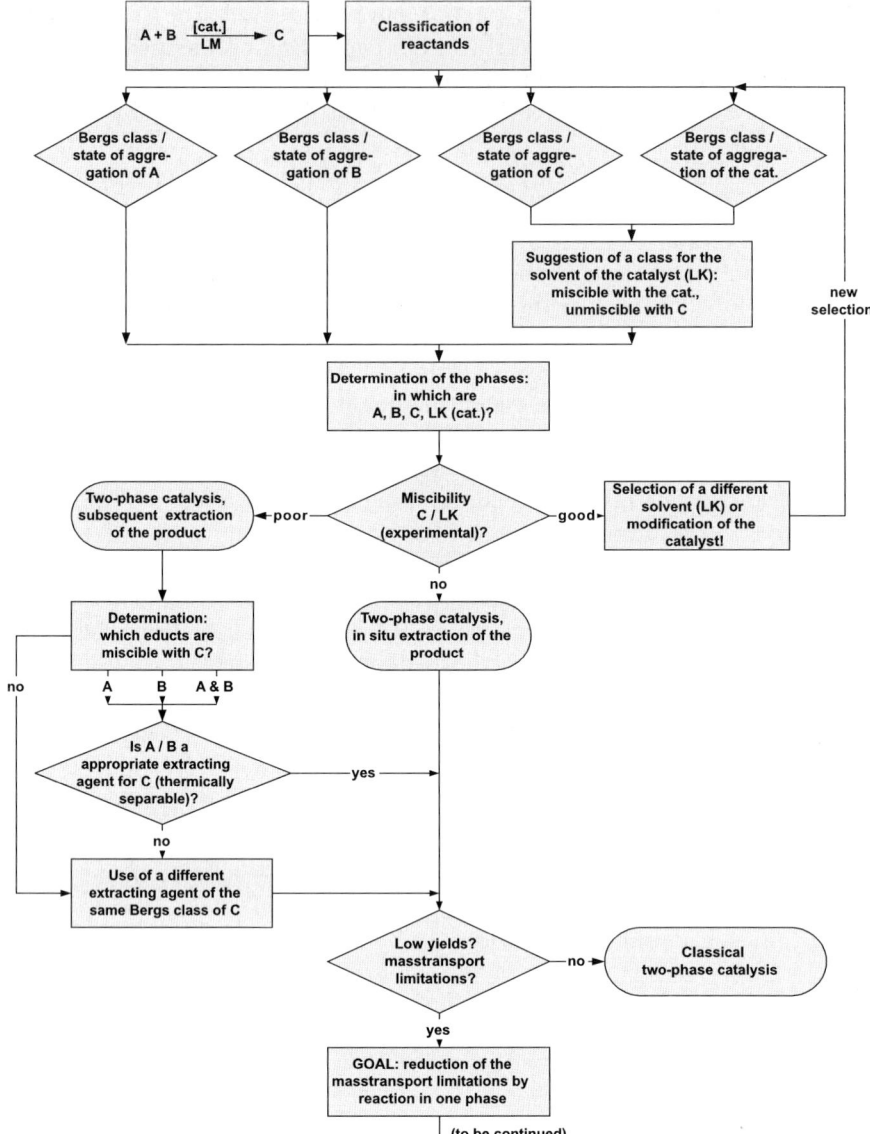

Fig. 15 "Decision tree" for solvent selection

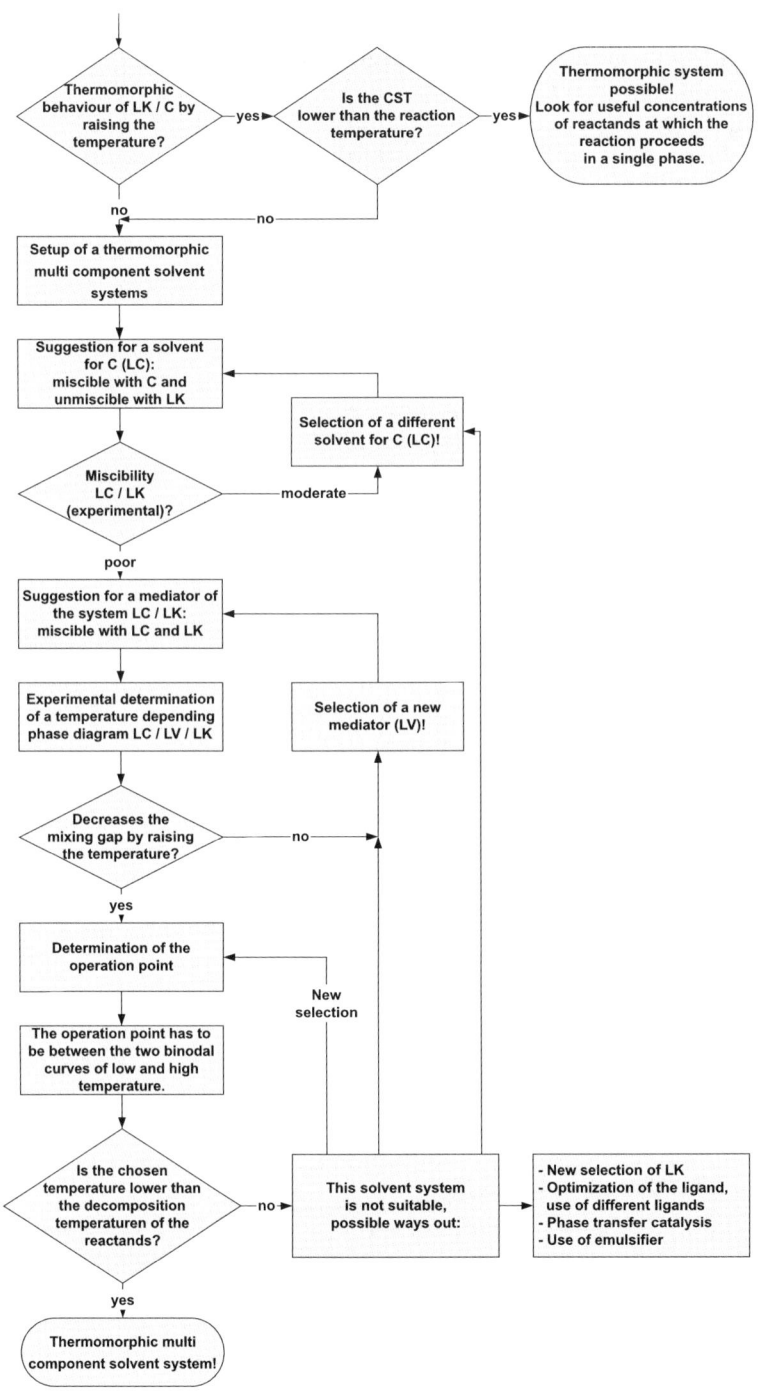

Fig. 15 (continued)

2.5
Development of an Expert System for Solvent Selection

To find systematically the right solvents for TMS systems it appeared desirable to establish a "decision tree" for solvent selection [41]. Subsequently this tree was implemented into an "expert system program" [42]. This program is based on the miscibility rules of Berg [43] who classified the solvents into five groups (Table 13).

This program guides the user through the "decision tree" and calculates the miscibility of the reactants. On the basis of these calculations and experimentally proved results a two phasic solvent system is developed. The miscibility of two substances is determined as follows:

- All substances of the same or an adjacent class (± 1) are completely miscible.
- Class 1 and 5 are immiscible.
- All other combinations are usually partially miscible and have to be determined experimentally.

If the reaction is limited by mass transfer, the program will assist the user to generate a TMS system to overcome this problem. A principle layout of the program is shown in Fig. 15.

3
Conclusions

The application of thermomorphic solvent systems as a new recycling concept was investigated in various C – C bond-forming reactions. Therefore methods for a systematic choice of solvent combinations were developed. In addition to common organic solvents more unusual solvents like cyclic carbonates, pyrrolidones, polyethylene glycols and lactones were used in the investigations. The phase behaviour of the new solvent systems was determined by cloud titrations. From these experiments information about the temperature dependency and an appropriate composition for the reactions could be obtained. The results were used in the development of an expert system for the solvent selection.

Suitable thermomorphic solvent systems were applied to the telomerization of butadiene with ethylene glycol or with carbon dioxide, to the isomerizing hydroformylation of *trans*-4-octene and to the hydroaminomethylation of 1-octene with morpholine. Further investigations for the carboxytelomerization and for the synthesis of 4-nitrodiphenylamine were also carried out. In addition to common ligands, PEG-modified ligands and fluorous ligands were used for the telomerization and the carboxytelomerization.

The use of TMS systems for the telomerization reactions did not lead to efficient reaction/separation processes until now. However, some solvent systems were determined, which could be used for other reactions. Basic principles of the influence of the substrates, products and the catalyst were investigated and the results were applied to further reactions.

If the polarity of the reaction mixture does not change too much during the reaction, much better results can be achieved: it could be shown that the isomerizing hydroformylation leads to very high conversions (99%) of the trans-4-octene and also to very attractive selectivities of n-nonanal ranging from 82 to 90% in the thermomorphic TMS-system PC/dodecane/p-xylene. To avoid the high rhodium leaching in this system, the mediator p-xylene was substituted by pyrrolidones. Especially with N-methylpyrrolidone, excellent conversions of 99% and high selectivities to n-nonanal of 81% are obtained. In this case less than 0.1% of the rhodium catalyst is lost to the product phase. The hydroaminomethylation of 1-octene with morpholine can also be performed in various thermomorphic solvent systems consisting of PC, an alkane and a semi-polar mediator. High conversions of 1-octene up to 98% after two hours at 125 °C and good selectivities to the amines of up to 96% can be achieved. Only a very low rhodium leaching to the product phase can be detected.

Acknowledgements We would like to thank Prof. Dr. Leitner (RWTH Aachen) for the supply of PEG-modified ligands and Prof. Dr. Gladysz (University of Erlangen-Nürnberg) and Prof. Dr. Dinjus (FZ Karlsruhe) for the donation of fluorous ligands.

We are very grateful to Umicore AG and Co. KG for the supply of the rhodium and palladium catalysts and to BASF AG for the donation of carbon monoxide and syngas. We also thank Celanese AG and European Oxo GmbH for the supply of TPPTS solution. We would like to thank the Bundesministerium für Bildung und Forschung (ConNeCat-project "Smart Solvents—Smart Ligands") and the "Fonds der Chemischen Industrie" for financial support.

References

1. Cornils B, Herrmann WA (2002) Applied Homogeneous Catalysis with Organo-metallic Compounds. Vol 1–3, 2nd edn. Wiley, Weinheim
2. Behr A, Toslu N (1999) Chem Ing Tech 71:490
3. Behr A, Toslu N (2000) Chem Eng Technol 23:122
4. Behr A, Fängewisch C (2001) Chem Ing Tech 73:874
5. Behr A, Fängewisch C (2002) Chem Eng Technol 25:143
6. Behr A, Urschey M (2003) J Mol Catal A: Chem 197:101
7. Behr A, Urschey M, Brehme VA (2003) Green Chemistry 5:198
8. Behr A, Urschey M (2003) Adv Synth Catal 345:1242
9. Behr A, Roll R (2005) Chem Ing Tech 77:748
10. Reichhardt C (2003) Solvents und Solvent Effects in Organic Chemistry. 3rd edn. Wiley, Weinheim, p 411
11. Tilloy S, Bertoux F, Mortreux A, Monflier E (1999) Catal Today 48:245

12. Monflier E, Fremy G, Castanet Y, Mortreux A (1995) Angew Chem Int Ed Eng 34:2269
13. Mathivet T, Méliet C, Castanet Y, Mortreux A, Caron L, Tilloy S, Monflier E (2001) J Mol Catal A: Chem 176:105
14. Monflier E, Bricout H, Hapiot F, Tilloy S, Aghmiz A, Masdeu-Bultó AM (2004) Adv Synth Catal 346:425
15. Behr A, Juszak KD, Keim W (1983) Synthesis 7:574
16. Behr A, Heite M (2000) Chem Eng Technol 23:952
17. Behr A, Heite M (2000) Chem Ing Tech 72:57
18. Behr A, Brehme VA (2002) J Mol Catal A: Chem 187:69
19. Behr A, Bahke P, Becker M (2004) Chem Ing Tech 76:1828
20. Weissermel K, Arpe HJ (2003) Industrial Organic Chemistry. 4th edn. Wiley, Weinheim
21. van Leeuwen PWNM (2000) Rhodium Catalyzed Hydroformylation. Kluwer Academic Publishers, Dordrecht
22. http://www.kitco.com/charts/rhodium.html, Kitco Inc.—Precious Metals, 2004
23. Behr A, Obst D, Schulte C, Schosser T (2003) J Mol Catal A: Chem 197:115
24. Behr A, Obst D, Turkowski B (2005) J Mol Catal A: Chem 226:215
25. Behr A, Henze G, Turkowski B, Obst D (2005) Green Chemistry 7:645
26. Hildebrand J, Scott RL (1950) The Solubility of Nonelectrolytes. 3rd edn. Reinhold, New York
27. Hildebrand J, Scott RL (1962) Regular Solutions. Prentice-Hall Inc., Englewood Cliffs, NJ
28. Hansen CM (1967) The Three Dimensional Solubility Parameter – Key to Paint Component Affinities I. J Paint Technol 39:104
29. Hansen CM (1967) The Three Dimensional Solubility Parameter – Key to Paint Component Affinities II–III. J Paint Technol 39:505
30. Hansen CM (2000) Hansen Solubility Parameters: a User's Handbook. CRC Press, Boca Raton, Florida
31. Barton AFM (1983) Handbook of Solubility Parameters and Other Cohesion Parameters. CRC Press, Boca Raton, Florida
32. Cheremisinoff NP (2003) Industrial Solvents Handbook, 2nd edn. Marcel Dekker, New York
33. Eilbracht P, Bärfacker L, Buss C, Hollmann C, Kitsos-Rzychon BE, Kranemann CL, Rische T, Roggenbuck R, Schmidt A (1999) Chem Rev 99:3329
34. Seayed A, Ahmed M, Klein H, Jackstell R, Gross T, Beller M (2002) Science 297:1676
35. Achmed M, Seayad A, Jackstell R, Beller M (2003) J Am Chem Soc 12:10311
36. Rische T, Eilbracht P (1997) Synthesis 11:1331
37. Rische T, Kitsos-Rzychon B, Eilbracht P (1998) Tetrahedron 54:2723
38. Rische T, Bärfacker L, Eilbracht P (1999) Eur J Org Chem 3:653
39. Behr A, Fiene M, Buss C, Eilbracht P (2000) Eur J Lipid Sci Technol 102:467
40. Behr A, Roll R (2005) J Mol Catal A: Chem 239:180
41. Behr A, Ott R, Turkowski B (2003) XXXVI. Jahrestreffen Deutscher Katalytiker, Weimar Tagungsband: 449
42. Behr A, Schöbel R, Roll R (2005) XXXVIII. Jahrestreffen Deutscher Katalytiker, Weimar Tagungsband: 381
43. Berg L, Harrison JM, Ewell RH (1944) Ind Eng Chem 36:871

Top Organomet Chem (2008) 23: 53–66
DOI 10.1007/3418_2006_061
© Springer-Verlag Berlin Heidelberg
Published online: 9 May 2007

PEG-Modified Ligands for Catalysis and Catalyst Recycling in Thermoregulated Systems

Ellen Hermanns · Jens Hasenjäger · Birgit Drießen-Hölscher[†]

Institut für Technische Chemie und Chemische Verfahrenstechnik,
Universität Paderborn, Warburger Str. 100, 33098 Paderborn, Germany

[†]Prof. Dr. B. Drießen-Hölscher deceased on November 16, 2004. Subsequently, the working group at the university of Paderborn was closed. Questions regarding this text can be adressed to Prof. Dr. Walter Leitner (leitner@itmc.rwth-aachen.de).

Abstract In this chapter recent developments in catalyst recycling employing thermomorphic ligands based on polyethylene glycol will be reviewed. The main focus will lie on the hydroformylation of higher olefins. The general principles of thermoregulated phase-transfer catalysis and thermoregulated phase-separable catalysis will be explained and the use of this recycling technique in hydroformylation catalysis exemplified. Different synthetic approaches toward novel polyethylene glycol modified monocyclic phosphites will be evaluated. The performance of new phosphite ligands in the rhodium-catalyzed hydroformylation of 1-octene will be described, and structure–selectivity and structure–activity dependencies will be discussed.

Keywords Immobilization · Thermoregulated catalysis · Thermomorphic ligands · Hydroformylation · Polyethylene glycol modified phosphites

Abbreviations
Cat	Catalyst
Cp	Cloud point
CST	Critical solution temperature
MME	Monomethyl ether
P	Product

PEG   Polyethylene glycol
S   Substrate
TOF   Turnover frequency
TRPSC   Thermoregulated phase-separable catalysis
TRPTC   Thermoregulated phase-transfer catalysis

1
Thermoregulated Catalysis

The immobilization of molecular catalysts is of great current interest [1], as it attempts to combine the positive aspects of homogeneous and heterogeneous catalysis [2]. A very good example of an immobilized catalyst system is the aqueous biphasic hydroformylation developed by Ruhrchemie/Rhône-Poulenc [3, 4]. As discussed in other chapters of this volume, these novel approaches include biphasic solvent systems based on fluorocarbons [5, 6], supercritical CO_2 [7], ionic liquids [8], and special mixtures of organic solvents [9]. The present chapter will focus on the immobilization of hydroformylation catalysts by using thermomorphic systems based on ligands that are modified with polyethylene glycol (PEG) side chains. This concept takes advantage of the temperature-dependent solubility of polymer-supported ligands in water and organic solvents. Two types of thermoregulated catalysis, thermoregulated phase-transfer catalysis and thermoregulated phase-separable catalysis, and their application to hydroformylation are described below.

1.1
Thermoregulated Phase-Transfer Catalysis

The water solubility of surfactants with polyoxyethylene moieties as the hydrophilic groups is based on hydrogen bonds formed between polyether chains and water molecules. The solubility of this class of surfactants decreases with an increase in temperature. Therefore, the aqueous solutions of this type of surfactants undergo an interesting phase separation process on heating to the cloud point, where the loss of hydrogen bonding leads to a miscibility gap. This behavior is exploited in thermoregulated phase-transfer catalysis (TRPTC) (Fig. 1) [10].

The principle of this process relies on the fact that at temperatures lower than the cloud point, the catalyst remains in the aqueous phase. On heating to temperatures above the cloud point, however, the catalyst is transferred into the organic phase. Thus, the catalyst and the substrate are in the same phase and the homogeneously catalyzed reaction can proceed. As soon as the reaction is completed and the system is cooled to a temperature below the cloud point, the catalyst returns to the aqueous phase and can be recycled.

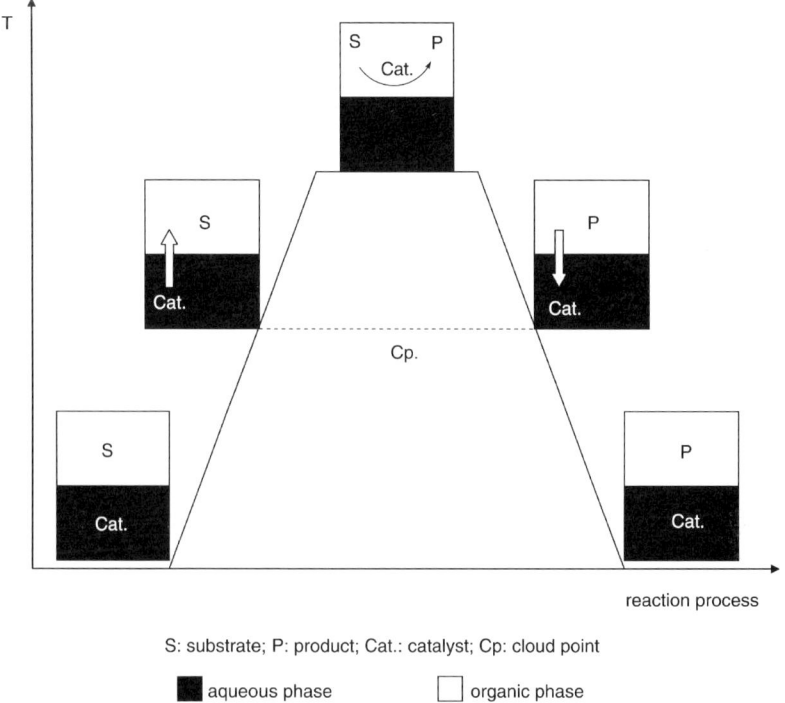

S: substrate; P: product; Cat.: catalyst; Cp: cloud point

■ aqueous phase   □ organic phase

Fig. 1 The general concept of thermoregulated phase-transfer catalysis (TRPTC)

A series of water-soluble polyether-substituted triphenyl phosphines (PETPPs) **1a–c** has been successfully employed by Jin et al. [11] in the thermoregulated hydroformylation of 1-dodecene in the biphasic water/toluene system. The catalysts exhibit very good catalytic properties with conversions up to 93% and about 85% selectivity for aldehyde formation. The catalyst derived from rhodium(III) chloride and ligand **1c** could be reused in four consecutive cycles without significant loss of activity or chemoselectivity. The *n*-selectivity of the product aldehydes was not determined.

The use of catechol-based phosphites with PEG moieties in the thermoregulated phase-transfer hydroformylation of 1-decene was also investi-

n x m = 8-25
m = 1, 2, 3
1 a, b, c

gated [12]. Utilizing **2a,b**/Rh catalysts formed in situ, conversions of > 90% of 1-decene and high aldehyde yields were obtained, but more branched than linear aldehydes were formed. The catalysts could be separated after the reaction by simple decantation, but with considerable loss in activity in the consecutive runs due to hydrolysis of the phosphite.

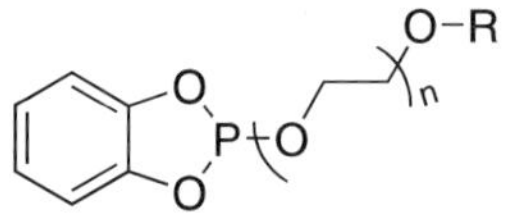

$$R = C_8H_{17}, C_6H_5; n = 8\text{-}66$$

2 a, b

In 2003 Jin and coworkers [13] reported on the synthesis of a novel nonionic water-soluble phosphine **3** by a two-step ethoxylation of 2-(diphenylphosphino)phenylamine. In this new family of nonionic phosphines the polyoxyethylene moieties are introduced into the amino group of organophosphines.

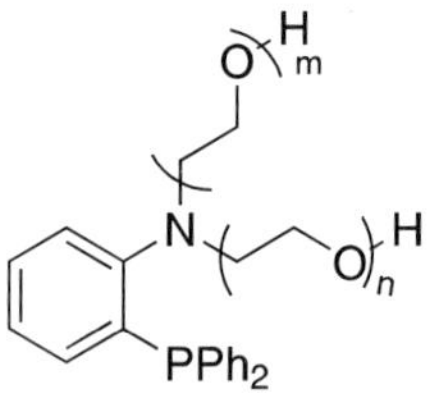

$$m + n = 35, 45$$

3

The **3**/Rh catalyst formed in situ (P/Rh = 4 : 1, substrate/Rh = 1000 : 1) has been applied to the aqueous–organic biphasic hydroformylation of 1-decene. The conversion of olefin and yield of aldehyde were 99.5 and 99.0%, respectively, after 5 h at 120 °C and 5.0 MPa of synthesis gas. Recycling tests showed that the aldehyde yield was still higher than 94.0% even after the catalyst had been recycled 20 times.

Octylpolyglycol phenylene phosphite **4** was investigated as ligand in the thermoregulated phase-transfer hydroformylation of styrene [14]. The cata-

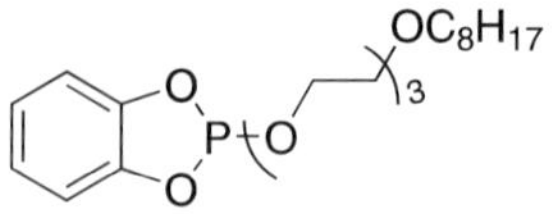

4

lyst displayed excellent catalytic activity (99.6% conversion and 99.3% aldehyde yield) in water/*n*-heptane under optimized conditions. The selectivity toward the branched aldehyde was 83%.

Lemaire and coworkers [15] published in 2000 the first enantioselective thermoregulated phase-transfer hydroformylation. Chiral PEG-modified phosphites 5 and 6 based on (S)-biphenol combined with a rhodium precursor exhibit high catalytic activity in the enantioselective hydroformylation of styrene. These thermoregulated phase-transfer catalysts gave branched aldehydes with high regioselectivity and *ee* of up to 25%, which is exactly the same value that has been obtained in homogeneous systems. Recycling was possible, but with a lower activity since rhodium leaching could not be avoided.

n = 16-17

5

n = 4-5

6

1.2
Thermoregulated Phase-Separable Catalysis

In 1993 Bergbreiter and coworkers [16] reported a catalyst system based on nonionic, water-soluble phosphine/rhodium catalysts which exhibited a temperature-dependent solubility behavior that led to precipitation (separation) rather than phase transfer as described above. These kinds of ligands were developed from block polymers of propylene oxide and ethylene oxide, which are water-soluble below and insoluble above the lower critical solution point. While the exact lower critical solution temperature varied with the proportion of polyethylene oxide/polypropylene oxide in the oligomer, a typical oligomer having a molecular weight of 2500 and 20 mol % was soluble at 0 °C in water, but insoluble above room temperature. In addition to the possibility of catalyst recycling, such catalysts include a "built-in" temperature control: they separate from a reaction mixture that warms up owing to an exothermic process, thus providing a molecularly determined shutoff temperature. Therefore, these ligands have been referred to as "smart ligands", an expression that has been extended to ligands with a controllable solubility behavior in more general terms.

The opposite temperature-dependent solubility of ligands in organic solvents is applied in the thermoregulated phase-separable catalysis (TRPSC) first published by Jin and coworkers [17] in 2000. The general principle is shown in Fig. 2 [10].

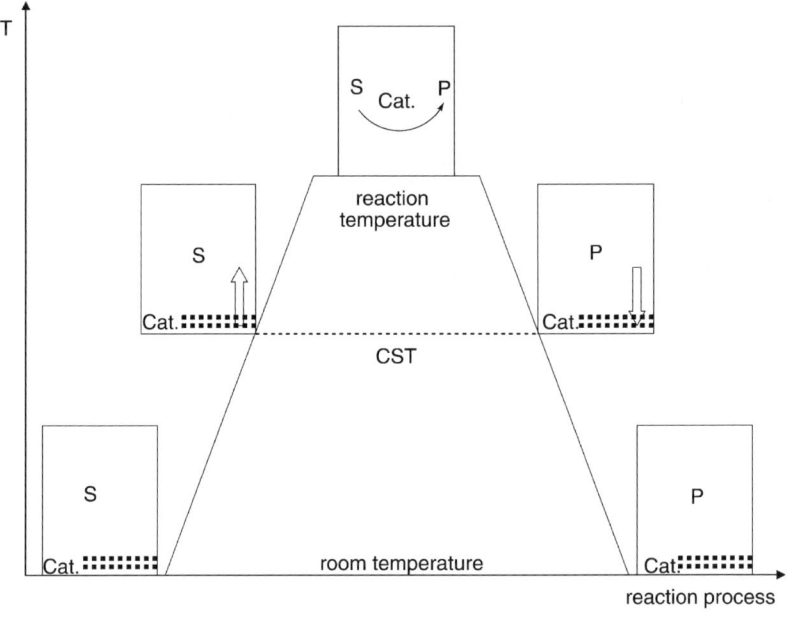

S: substrate; P: product; Cat. catalyst; CST: critical solution point

□ organic phase

Fig. 2 The general concept of thermoregulated phase-separable catalysis (TRPSC)

Before the reaction, at temperatures below the critical solution point, the catalyst is insoluble in the organic solvent. When heated to temperatures above the critical solution point, the catalyst is soluble in the organic solvent and a homogeneous system is formed in which the catalyzed reaction takes place. After reaction, on cooling to temperatures below the critical solution point, the catalyst precipitates from the organic phase which contains the product. Thus, the catalyst can be easily separated from the product by decantation or filtration and reused.

The nonionic phosphine 7 was used in the rhodium-catalyzed hydroformylation of 1-dodecene under thermoregulated phase-separable conditions. Phosphine 7 is soluble in toluene at higher temperatures, but can be

$$n = 6$$

7

precipitated after reaction by a decrease in temperature. This system leads to 96% conversion and 94% aldehyde yield. Eight consecutive runs were possible without significant loss of activity [17].

Liu et al. [18] investigated the possibility of catalyst recycling in the nonaqueous hydroformylation of 1-decene by using the thermomorphic polyether phosphite **2a** described earlier under phase-transfer conditions. Catalyst recovery with the procedure of phase-separable catalysis was possible with 0.92% rhodium loss in the seventh cycle. Complete olefin conversion and aldehyde yields of 98% were reached, but linear and branched aldehydes were formed in almost equal amounts.

2
Novel Polyethylene Glycol Modified Phosphites

2.1
Synthesis

2.1.1
Synthetic Strategy

Ligands with PEG side chains are of interest for thermoregulated catalysis owing to their temperature-dependent solubility. In monocyclic phosphites the PEG moiety can be bound directly to the phosphorus atom. Besides catechol-based phosphites, in analogy to Jin and coworkers [12, 14, 18] phosphites adapted from 1,1-biphenol and 5,5′-dichloro-6,6′-dimethoxybiphenyl-2,2′-diol (ClMeO-biphenol) were synthesized by our group [19, 20] within the ConNeCat network. Two different synthetic approaches toward PEG-modified monocyclic phosphites are possible as shown in Scheme 1. On the one hand, the diol **8** can first be transformed into the corresponding phosphorochloridite **9** and then reacted with the PEG alcohol **10** to give phosphite **11**. On

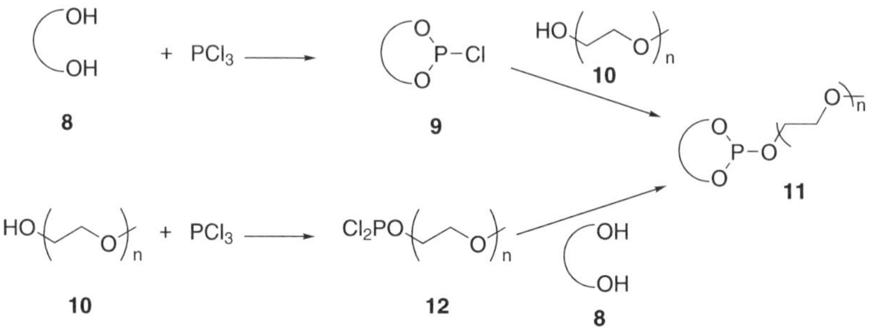

Scheme 1 Synthetic approaches toward PEG-modified monocyclic phosphites

the other hand, the PEG alcohol **10** can first be transformed into the phosphorochloridite **12** and afterward the corresponding phosphite **11** is formed on addition of diol **8**.

2.1.2
Catechol-Based Phosphites

The mono- to tetraethylene glycol modified phosphites **14a–d** have been synthesized according to the first-described method starting from commercially available (*o*-phenylene) phosphorochloridite **13** in > 90% yield (Scheme 2).

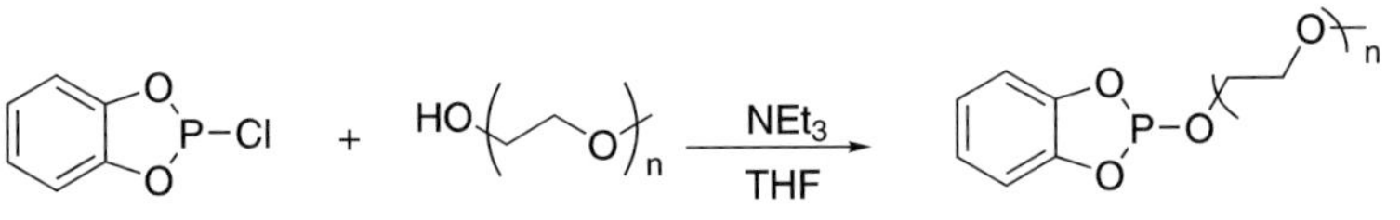

13　　　　　　　　　　**10 a, b, c, d**　　　　　　　　**14　a, b, c, d**
　　　　　　　　　　　　　　n = 1, 2, 3, 4　　　　　　　　　　　n = 1, 2, 3, 4

Scheme 2 Synthesis of PEG-modified catechol-based phosphites

In the present study, a series of ligands was targeted comprising different cyclic backbones on one side and well-defined short oligomeric ether chains on the other side. This set of ligands should help to identify structural influences on the catalytic performance of these systems. Although these ligands do not show thermomorphic properties by themselves, the findings can provide guidelines for the selection of the higher molecular weight derivatives. In contrast to higher molecular weight PEGs, which are available with a certain distribution only, the short-chain ligands can be obtained with well-defined structures and are hence preferred for systematic evaluation.

2.1.3
Phosphites Based on the Biphenol Backbone

A second set of ligands was prepared on the basis of biphenol as backbone for the cyclic structure. This results in a seven-membered rather than a five-membered ring as compared to the catechol-based phosphites. Furthermore, the biphenol derivatives can be expected to be sterically more demanding than catechol ligands.

1,1′-Biphenyl-2,2′-diylphosphorochloridite (**16**) was synthesized according to Cuny and Buchwald [21] and purified via flash distillation. Reaction of **16** and mono- to hexaethylene glycol monomethyl ethers **10a–e** yielded the corresponding phosphites **17a–e** in good yields.

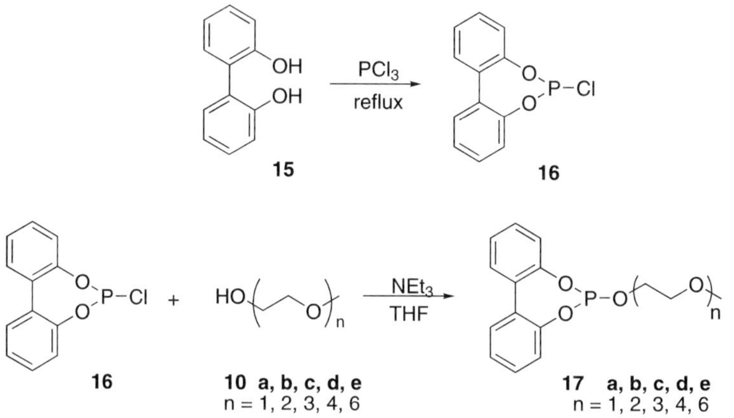

Scheme 3 Synthesis of PEG-modified phosphites **17**

2.1.4
Phosphites Based on the ClMeO-biphenol Backbone

As discussed for biphenol-based phosphites, phosphites derived from ClMeO-biphenol should also be sterically more demanding and have an increased ring size as compared to catechol-based phosphites. Moreover, by comparing the catalytic results of biphenol- and ClMeO-biphenol-based ligands, conclusions on the influence of the substitution pattern and resulting electronic effects on the activity and selectivity in the hydroformylation can be drawn. Therefore, ClMeO-biphenol-based phosphites with PEG side chains consisting of one to three ethylene glycol units were included in the present study.

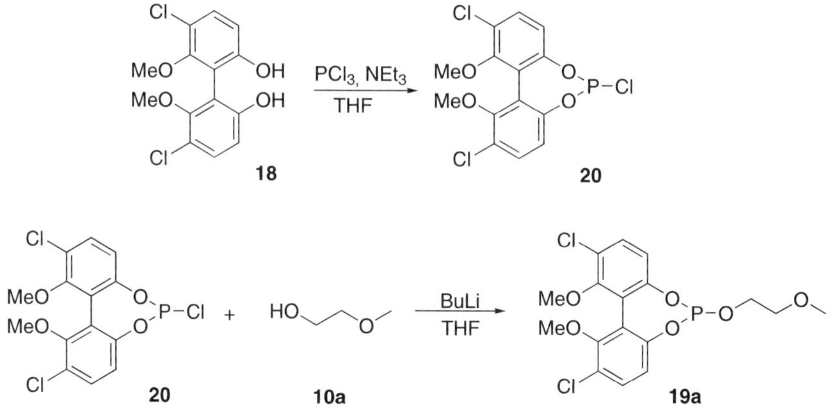

Scheme 4 Synthesis of ClMeO-biphenol-based phosphite **19a**

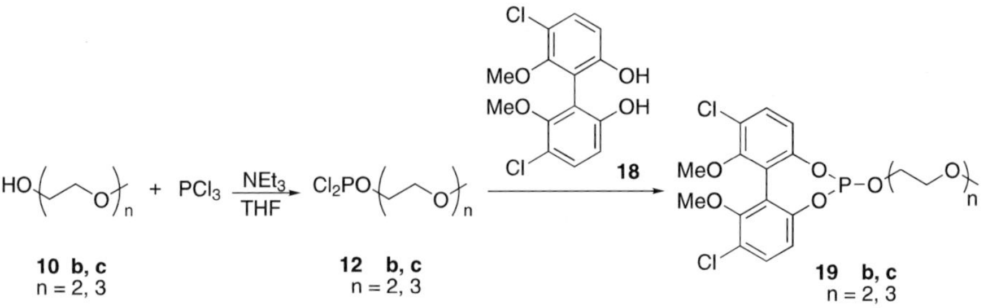

Scheme 5 One-pot synthesis of ClMeO-biphenol-based phosphites **19b,c**

2-Methoxyethanol-modified phosphite **19a** was prepared in 87% yield [19] according to the two-step procedure for the synthesis of enantiomerically pure phosphites with the ClMeO-biphenol backbone developed in our group [22–24] (Scheme 4). The ligands **19b,c** with slightly longer ether chains were prepared following the procedure described recently by Huang et al. for BINOL-based phosphites (Scheme 5) [25]. Yields were slightly lower (80%), but conducting the couplings in a one-pot procedure makes this protocol very attractive.

2.2
Catalytic Performance

In order to assess structural influences on the catalytic performance of monocyclic PEG-modified phosphites, the monocyclic phosphites **14a,c**, **17a–e**, and **19b,c** were used in the rhodium-catalyzed hydroformylation of 1-octene (Scheme 6). The catalysts were prepared in situ by adding a solution of [Rh(acac)(CO)$_2$] as a catalyst precursor in toluene to the phosphite ligands

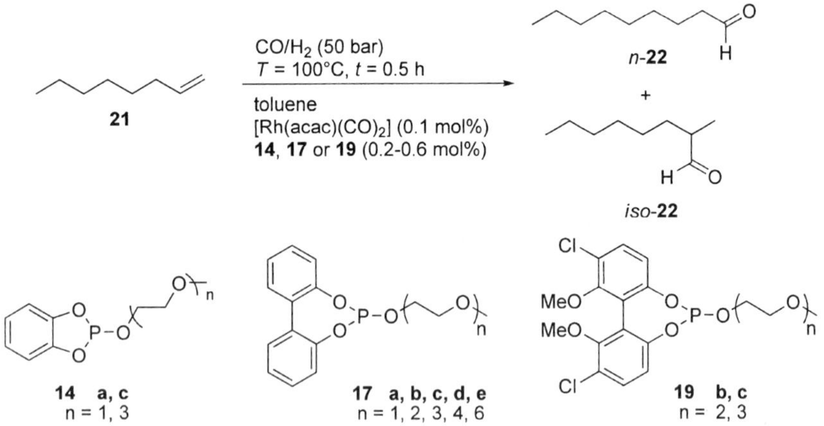

Scheme 6 Rh-catalyzed hydroformylation of *n*-octene using phosphites **14**, **17**, and **19**

dissolved in the same solvent. Preformation under hydroformylation conditions in the absence of the olefin led to the active hydroformylation catalysts. The reactions were started by adding the olefin to the catalyst solution. The preformation and reaction were conducted in stainless steel high-pressure reactors equipped with an overhead stirrer and a reservoir for the olefin.

All ligands were employed in the hydroformylation of 1-octene (21) under standard reaction conditions ($T = 100\,^\circ\mathrm{C}$, initial pressure (at $25\,^\circ\mathrm{C}$) 50 bar ($CO/H_2 = 1:1$), 0.04 mmol [Rh(acac)(CO)$_2$], 10 mL toluene, 1-octene/Rh = 1000, ligand/Rh = 2, 0.5 h preformation, 0.5 h reaction). The results of this ligand screening are summarized in Table 1. The data are given as conversion of 21, turnover frequency (TOF) for the formation of aldehydes 22, and regioselectivity toward the linear aldehyde n-22. The main side reaction was isomerization to the internal olefins that were not fully hydroformylated under the present conditions. The influence of the phosphite ligands is clearly reflected in the comparison of the individual results with that of an unmodified rhodium catalyst (entry 10). All catalysts lead to almost quantitative conversion of substrate 21. There are, however, significant differences in the chemoselectivity toward the aldehydes and the regioselectivity for the linear product n-22.

The catechol-based ligands 14 show fairly low selectivity similar to that of the unmodified system. In contrast, the biphenol-derived phosphites 17 gave excellent chemoselectivity. The best results were obtained with 17c, where a TOF of ca. $1900\,\mathrm{h^{-1}}$ was achieved which is close to the maximum value of $2000\,\mathrm{h^{-1}}$. The n-selectivity of 71% was also appreciable, corresponding to an n/iso ratio of 2.4. The substituents in the biphenol backbones in ligands 19 re-

Table 1 Hydroformylation of 1-octene using PEG-modified monocyclic phosphite ligands[a]

Entry	Ligand	Conversion (%)	n-selectivity[c]	TOF (h^{-1})[b] (%)
1	14a	97	63	832
2	14c	97	59	1080
3	17a	98	67	1853
4	17b	99	69	1851
5	17c	99	71	1913
6	17d	99	70	1843
7	17e	99	68	1717
8	19b	98	66	1652
9	19c	99	65	1600
10	–	98	57	1011

[a] Standard conditions
[b] Turnover frequency (TOF) = mol aldehyde × (mol rhodium)$^{-1}$ × h^{-1}
[c] $[n\text{-}22/(n\text{-}22 + iso\text{-}22)] \times 100$

duce both the chemo- and regioselectivity slightly. The length of the polyether chain also affects the selectivity slightly. In the series of ligands **17a** to **17e**, the best performance was observed with ligand **17c** containing three ethylene glycol units. Ligand **17e** with six units showed a slightly poorer performance. Further studies are required to investigate the generality of this trend, as this would be a possible limitation of the thermoregulated ligand systems which require PEG groups of even longer chain lengths ($n \geq 15$).

The influence of the ligand to metal ratio was assessed for the most successful ligands (Table 2). In the case of the biphenol-derived ligands **17**, the activity of the catalytic system decreased by at least 50% upon increasing the P/Rh ratio from 2 : 1 to 6 : 1. At the same time, the regioselectivity increased to 77% for **17c**. In contrast, both aldehyde formation and regioselectivity increased with increasing P/Rh for ligand **17c**. Excellent performance with 98% conversion, a TOF close to $1900\ h^{-1}$ for the formation of **22**, and an *n/iso* ratio of 3 : 1 was observed under the conditions of Table 2, entry 6.

The different response to changes in the P/Rh ratio for ligands **17c** and **19c** may reflect their different coordination ability. Ligand **17c** appears to bind readily to rhodium and shows the typical behavior expected for ligands that exhibit an equilibrium between species containing two or three phosphorus ligands at the rhodium center. In contrast, **19c** seems to bind less effectively and a higher excess of ligand is required to form the desired double ligated species.

Table 2 Influence of the ligand concentration on the selectivity and activity of the hydroformylation of 1-octene[a]

Entry	Ligand	P/Rh	Conversion (%)	*n*-Selectivity (%)[c]	TOF (h^{-1})[b]
1	**17a**	2	98	67	1853
2	**17a**	6	46	76	783
3	**17c**	2	99	71	1913
4	**17c**	6	53	77	933
5	**19c**	2	99	65	1600
6	**19c**	6	98	75	1866

[a] Reaction conditions: $T = 100\ ^\circ C$, initial pressure (at $25\ ^\circ C$) 50 bar ($CO/H_2 = 1:1$), 0.04 mmol [$Rh(acac)(CO)_2$], 10 mL toluene, 1-octene/Rh = 1000, 0.5 h preformation, 0.5 h reaction
[b] TOF = mol aldehyde $\times$ (mol rhodium)$^{-1} \times h^{-1}$
[c] $[n\text{-}\mathbf{22}/(n\text{-}\mathbf{22} + iso\text{-}\mathbf{22})] \times 100$

3
Conclusion

Thermoregulated phase-transfer and phase-separable catalysis are attractive catalyst recycling techniques complementing other approaches of multiphase catalysis. They utilize temperature-dependent solubility or miscibility phenomena to "switch" between homogeneous reaction and heterogeneous separation stages.

In most cases, a suitable molecular modification of the catalyst structure is required to obtain the desired thermoresponsive properties. Polyether and in particular PEG substituents are receiving considerable interest in this field. The present study has addressed structure–activity relationships for well-defined low molecular weight model ligands in the rhodium-catalyzed hydroformylation of 1-octene as benchmark reaction. Figure 3 summarizes the observed trends.

Promising performances were achieved with ligands **17** at the ratio P/Rh = 2 : 1 and with ligands **19** at P/Rh = 6 : 1. Interestingly, maxima in hydroformylation activity and regioselectivity were observed for ligand **17c** with an intermediate chain length of three ethylene glycol units. Further studies are necessary to investigate the influence of the PEG chain more systematically, as this is decisive for the thermoregulated behavior of these systems. Recent findings from Bönnemann and Jin on the formation and stabilization of catalytically active rhodium nanoparticles in the presence of ligands with long PEG chains have to be taken into account in any mechanistic interpretations as well.

Only a detailed understanding of the subtle interplay between ligand structure and catalyst performance on the one hand and recyclability on the other

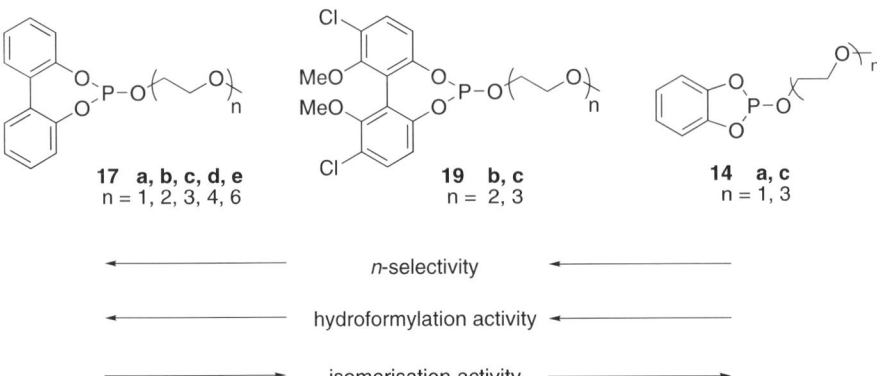

Fig. 3 General trend of activity and *n*-selectivity in the hydroformylation of 1-octene depending on the ligand backbone

will allow the design of effective and practical methods based on thermoregulated catalytic systems.

References

1. Association of the German Chemical Industry (VCI) (2002) Position paper about the further development in catalysis research in Germany (www.vci.de/fonds)
2. Hermann WA, Cornils B (1997) Angew Chem Int Ed Engl 36:1049
3. Kuntz E (1987) Chemtech 17:570
4. Kuntz E (1977) FR Patent 2 314 910
5. Horváth IT, Rabai J (1994) Science 266:72
6. Klement I, Lütjens H, Knochel P (1997) Angew Chem Int Ed Engl 36:1454
7. Jessop PG, Leitner W (eds) (1999) Chemical synthesis using supercritical fluids. Wiley, Weinheim
8. Wasserscheid P, Welton T (eds) (2004) Ionic liquids in synthesis. Wiley, Weinheim
9. da Rosa RG, Martinelli L, da Silva LHM, Loh W (2000) Chem Commun 33
10. Wang Y, Jiang J, Jin Z (2004) Catal Surv Asia 8:119
11. Jin Z, Zheng X, Fell B (1997) J Mol Catal A Chem 116:55
12. Liu X, Wang Y, Kong FZ, Jin Z (2003) Chin J Chem 21:494
13. Liu C, Jiang J, Wang Y, Cheng F, Jin Z (2003) J Mol Catal A Chem 198:23
14. Chen R, Liu X, Jin Z (1998) J Organomet Chem 571:201
15. Breuzard JAJ, Tommasino ML, Bonnet MC, Lemaire M (2000) J Organomet Chem 616:37
16. Bergbreiter DE, Zhang L, Mariagnanam VM (1993) J Am Chem Soc 115:9295
17. Wang Y, Jiang J, Zhang R, Liu X, Jin Z (2000) J Mol Catal A Chem 157:111
18. Liu X, Li H, Wang Y, Jin Z (2002) J Organomet Chem 654:83
19. Hasenjäger J (2002) Diploma thesis, RWTH Aachen
20. Hermanns E (2006) PhD thesis, RWTH Aachen
21. Cuny GD, Buchwald SL (1993) J Am Chem Soc 115:2066
22. Agel F, Drießen-Hölscher B, Dreisbach C, Prinz T, Militzer HC, Meseguer B, Scholz U (2003) EU Patent 1 298 136
23. Drießen-Hölscher B, Kralik J, Agel F, Steffens C, Hu C (2004) Adv Synth Catal 346:979
24. Agel F (2004) PhD thesis, RWTH Aachen
25. Huang H, Zheng Z, Luo H, Bai C, Hu X, Chen H (2004) J Org Chem 69:2355
26. Wen F, Bönnemann H, Jiang J, Lu D, Wang Y, Jin Z (2005) Appl Organomet Chem 19:81

Top Organomet Chem (2008) 23: 67–89
DOI 10.1007/3418_043
© Springer-Verlag Berlin Heidelberg
Published online: 29 July 2006

Temperature-Controlled Catalyst Recycling: New Protocols Based upon Temperature-Dependent Solubilities of Fluorous Compounds and Solid/Liquid Phase Separations

John A. Gladysz (✉) · Verona Tesevic

Institut für Organische Chemie, Friedrich-Alexander-Universität Erlangen-Nürnberg,
Henkestraße 42, 91054 Erlangen, Germany
gladysz@chemie.uni-erlangen.de

Abstract The absolute solubilities of fluorous compounds can be tailored by varying the lengths of the $(CF_2)_{n-1}CF_3$ (R_{fn}) segments. Many such compounds exhibit immense solubility increases in organic solvents or neat liquid reactants upon heating. Suitably designed fluorous catalysts can therefore be employed under homogeneous conditions at elevated temperatures, and recovered by solid/liquid phase separation at lower temperatures. Expensive fluorous solvents are avoided. Fluorous supports can be used to aid the recovery of small catalyst quantities, and render phase separation more efficient. Other design considerations, such as the nature of the catalyst rest state, are analyzed. Examples from the authors' laboratory arising from a multi-investigator project on multiphase catalysis are emphasized. These include phosphine-catalyzed additions, metallacycle syntheses and reactions, and rhodium-catalyzed hydrosilylations. Results from other laboratories are briefly described.

Keywords Fluorous · Hydrosilylation · Palladacycles · Phosphines · Recycling · Rhodium · Teflon® · Thermomorphic

1
Introduction

Following the publication of the first example of fluorous biphase catalysis by Horváth and Rábai in 1994 [1], the immediate focus was to develop
catalysts that would exhibit very biased partition coefficients with respect
to fluorous and organic solvents. Such liquids are normally immiscible at
room temperature. This was done by attaching "ponytails" of the formula
$(CH_2)_m(CF_2)_{n-1}CF_3$ (abbreviated $(CH_2)_m R_{fn}$), including arrays emanating
from silicon atoms [2]. Catalysis was then effected at elevated temperatures,
where fluorous and organic solvents are commonly miscible, with prod-
uct/catalysis separation at the low-temperature two-phase limit.

In this initial burst of activity, Hughes synthesized the fluorous ferrocenes
$(\eta^5\text{-}C_5H_4(CH_2)_2R_{fn})_2Fe$ (1-R_{fn}; $n = 6, 8, 10$), and commented on their solubilities [3]. The complex with the shortest ponytail, 1-R_{f6}, dissolved readily
in common nonpolar organic and fluorous solvents. Complex 1-R_{f8} was less
soluble, but appreciable concentrations could still be achieved. The complex
with the longest ponytail, 1-R_{f10}, "dissolved easily in diethyl ether, moderately in hexanes, and sparingly in chloroform, acetone, and toluene at room
temperature, although it dissolved easily in these solvents hot". A 0.1 M solution of 1-R_{f10} in $CF_3C_6F_{11}$ (perfluoro(methylcyclohexane)) could be realized
at 40 °C, but only a 0.003 M solution at room temperature. This represents
a ca. 33-fold concentration increase over less than 20 °C. The corresponding
solubilities of 1-R_{f8} and 1-R_{f6} at room temperature were 0.06 and > 0.09 M.

The solubilities of other fluorous compounds were subsequently noted to
greatly depend upon the lengths of the perfluoroalkyl segments. A few researchers remarked on the highly temperature-dependent solubilities, which
in at least some cases were discovered during efforts to acquire NMR spectra.
However, extended discourses on solubility properties do not normally make
for gripping reading, and are avoided by many authors. Therefore, wide general recognition of these properties remained "off the radar screen", delaying
possible applications.

$(CH_2)_2R_{fn}$

Fe

$(CH_2)_2R_{fn}$ $P((CH_2)_m R_{fn})_3$

1-R_{fn}[3]   5[36]

$n = 6, 8, 10$ $m/n = 2/8$, **a**

$R_{fn} = (CF_2)_{n-1}CF_3$ $m/n = 3/8$, **b**

Concurrently with the preceding developments, general interest in reducing solvent usage in synthesis was increasing. With fluorous/organic liquid/liquid biphase chemistry, the expense and environmental persistence of the fluorous solvent also pose disadvantages [4]. One day, during the doctoral studies of Marc Wende, Christian Rocaboy, and Long Dinh, it occurred to us that the

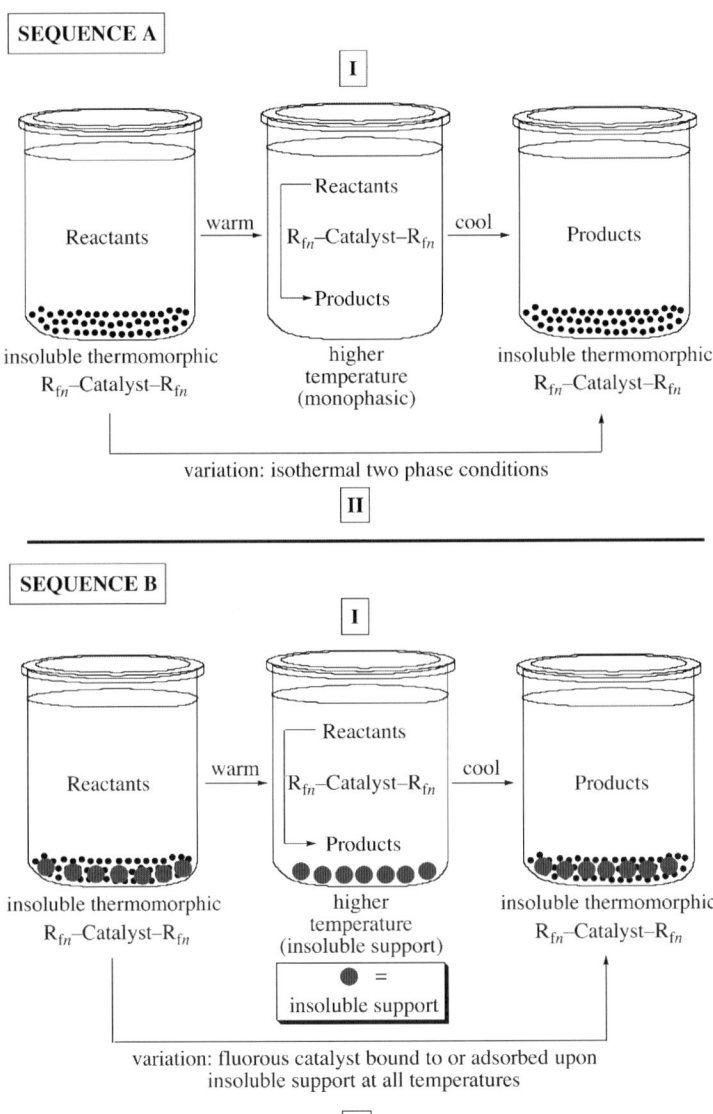

Fig. 1 Schematic of title processes: strategies for the recovery of fluorous catalysts via solid/liquid phase separation

temperature-dependent solubilities of fluorous molecules might be exploitable for recyclable catalysts that could be used under homogeneous conditions. In other words, one would make use of the temperature-dependent solubility of a solid in a liquid phase instead of the temperature-dependent miscibility of two liquid phases. Catalyst recovery would then be effected by a solid/liquid phase separation, as sketched in sequence A-I of Fig. 1.

The primary goal of this account is to summarize our efforts in this area, particularly in the context of a multi-investigator block grant for multiphase catalysis that was subsequently awarded and involved many of the other authors in this monograph. Additional researchers have also made important contributions to this area, and highlights of their results will be briefly treated below [5–12].

2
General Considerations

A compound with a highly temperature-dependent property, such as solubility, is said to be thermomorphic. There are a variety of other catalyst systems that can be recovered on the basis of temperature-dependent solubilities, usually involving polymers [13–18]. Attention has also been given to nonthermal "solubility switches". These include photochemically and chemically triggered "precipitons" [19–25], and the use of CO_2 pressure to regulate solubilities [26–28]. The latter protocol is particularly suited to fluorous compounds.

Any catalyst to be recovered on the basis of solubility should be designed to have as low a solubility as possible under the separation conditions. This will determine the lower bound for leaching. If quantitative solubility data are available, the theoretical leaching level can be calculated. A high solubility gradient with temperature is desirable, but not in our opinion essential. Although a higher catalyst concentration commonly leads to a faster rate, an alternative solution in the case of a poorly soluble catalyst is to design a more active catalyst. Carrying this point one step further, one can question whether any significant solubility is needed at all. If the reaction rate is sufficiently rapid in the presence of the solid catalyst, achieving homogeneous conditions is not a concern. This limit is represented by sequence A-II in Fig. 1. Indeed, several fluorous catalysts operate satisfactorily under apparently heterogeneous conditions [29]. There are, however, additional factors such as mass transfer and catalyst morphology that need to be considered under such conditions.

For every reaction, it is also important to consider *what is actually being recycled*. This is always the catalyst rest state, which is not necessarily the catalyst precursor. Particularly in the case of transition metal-based systems, irreversible steps are often necessary for the catalyst precursor to enter the catalyst cycle (e.g., ligand dissociation). Furthermore, the rest state may depend upon the reactant in excess, and some systems will have a distribution of rest states. In any event, the solubility properties of the rest state, and not the catalyst precursor, will determine the feasibility of sequence A-I in Fig. 1.

Thus, the solubility characteristics of the catalyst precursor may provide only a rough guide to those of the rest state. Accordingly, rational design can carry one only so far. However, some classes of catalysts have a higher probability of "resting" in their isolable form. The most obvious example would be Brønsted catalysts. With Lewis acids and bases, the question is whether they prefer to rest in their "native" form, or as some adduct of the product, or a reactant that may be in excess. There are many additional factors critical for the evaluation and comparison of recyclable catalysts, and these have been reviewed elsewhere [30–32].

3
First Test System: Thermomorphic Lewis Base Catalysis [33,34]

As shown in Scheme 1, aliphatic phosphines such as $P(n\text{-Bu})_3$ catalyze the addition of alcohols (2) to methyl propiolate (3) [35]. The mechanism is believed to involve an initial addition of the phosphine to the $C \equiv C$ moiety to give a zwitterionic allenolate (I), which then deprotonates the alcohol, yielding a vinyl phosphonium salt (II). An alkoxide addition to give an enolate (III), followed by phosphine elimination gives the product 4 and regenerates the catalyst. Several experiments suggest that when alcohols are used in excess, the catalyst rests as the original phosphine [34].

In earlier work, we prepared a series of fluorous aliphatic phosphines of the formula $P((CH_2)_m R_{fn})_3$ (5; $m = 2$–5; $n = 6, 8, 10$) [36, 37]. As with Hughes'

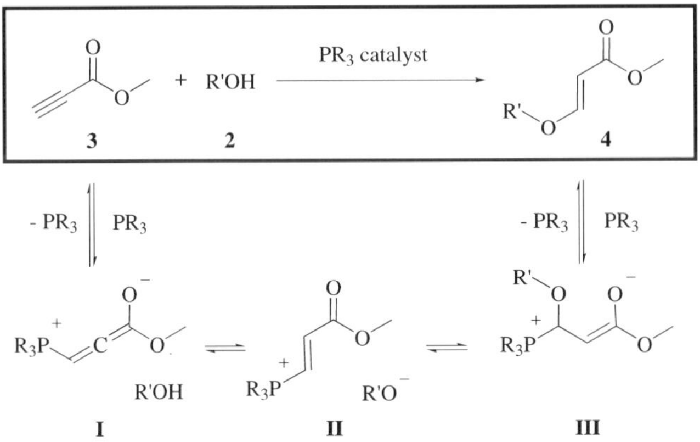

Scheme 1 Phosphine-catalyzed addition of alcohols (2) to methyl propiolate (3)

ferrocenes 1-R$_{fn}$, the solubilities were primarily determined by the lengths of the R$_{fn}$ segments. Data for **5a** ($m/n = 2/8$) in a spectrum of solvents are shown in Fig. 2. Data for **5b** ($m/n = 3/8$), which has one more insulating methylene group and should therefore be more basic and nucleophilic, were very similar [34].

Indeed, both phosphines were strongly thermomorphic, particularly with respect to the less polar solvent n-octane. Between 20 and 80 °C, the solubility of **5a** increased ca. 60-fold. Between 20 and 100 °C, the increase was 150-fold. More important were the low absolute concentrations at lower temperatures. Very little **5a** could be detected in n-octane at – 20 or 0 °C (0.104 and 0.308 mM). At 20 °C, millimolar concentration levels were present (1.13 mM). The low temperature limits were similar for the more polar solvents toluene and chlorobenzene (Fig. 2). Although the solubilities did not increase as much with temperature, note that those at 65 °C (≥ 6.5 mM) are sufficient for all of the catalytic reactions described below.

The effectiveness of **5a** as a catalyst for the addition of alcohols **2** to propiolate **3** was first established under conventional fluorous liquid/organic biphase conditions. This set the stage for the sequence in Fig. 3. Compounds **2a**, **3**, and **5a** were combined in n-octane at room temperature in a 2.0 : 1.0 : 0.1 ratio (10 mol % **5a**). As would be expected from Fig. 2, there was no visually

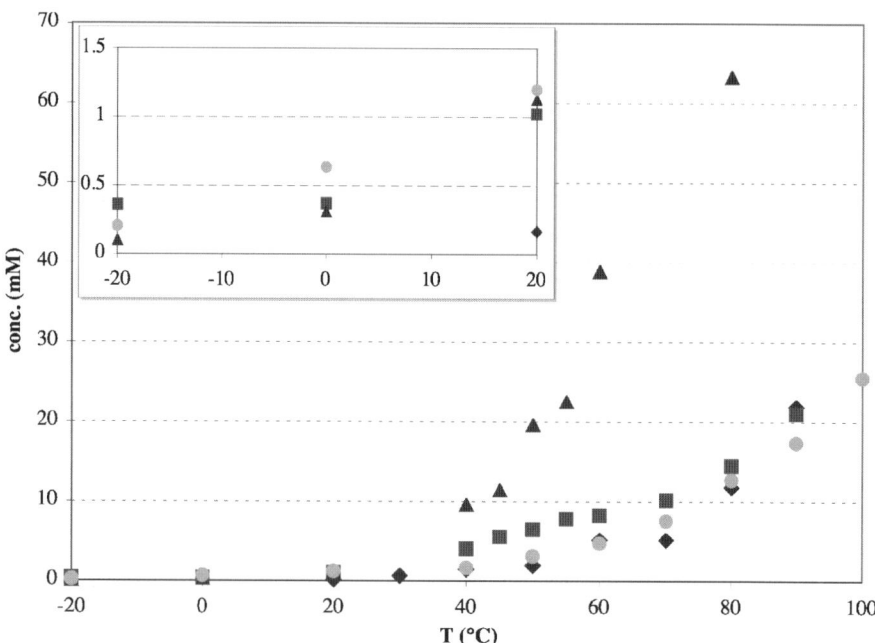

Fig. 2 Solubility of 5a in toluene (■), chlorobenzene •, dioxane (♦), and n-octane (▲) as a function of temperature

detectable dissolution of **5a**. The sample was warmed to 65 °C, and became homogeneous (ca. 6.5 mM in **5a**). After 8 h, the sample was cooled, and **5a** precipitated. In a few cases, the original white color was retained, but most samples were orange and some darkened to red with additional recycling. Photographs of representative sequences have been published [33, 34].

The supernatant was removed at – 30 °C, and the catalyst residue extracted once with cold *n*-octane. GC analysis indicated an 82% yield of **4a**. As summarized in Fig. 3 (entry 1), the recovered **5a** was used for four further cycles without deterioration in yield. Multiple runs could be similarly conducted with alcohols **2b–d**, affording comparable yields of **4b–d** (entries 2–4). A number of control experiments have been detailed elsewhere [33, 34]. Also, since **2a,c,d**, **3**, and **4a–d** are liquids at room temperature, these reactions

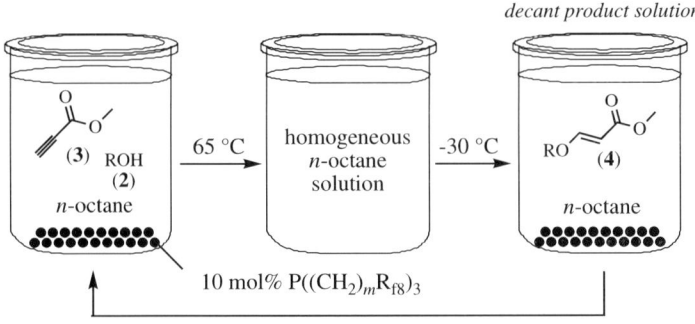

recycle solid $P((CH_2)_m R_{f8})_3$ (m = 2, **5a**; m = 3, **5b**)

R = **a**, $PhCH_2$; **b**, Ph_2CH; **c**, $PhCH(CH_3)$; **d**, $CH_3(CH_2)_7$

Entry	Reactants (2.0 : 1.0 mol ratio)	Catalyst (10 mol%)	Yield **4a-d** (%)				
			Cycle 1	Cycle 2	Cycle 3	Cycle 4	Cycle 5
1	**2a/3**	**5a**	82	82	80	81	75
2	**2b/3**	**5a**	77	84	71	-	-
3	**2c/3**	**5a**	90	86	75	-	-
4	**2d/3**	**5a**	79	84	66	-	-
5[a]	**2a/3**	**5a**	99	>99	97	95	-
6	**2a/3**	**5b**	81	89	82	80	77
7	**2c/3**	**5b**	92	97	99	-	-

[a] *n*-Octane solvent omitted.

Fig. 3 Recycling of thermomorphic fluorous phosphine catalysts **5a,b** via solid/liquid phase separations (Starting concentration of **2**, 1.25 M; cycle time, 8 h for **5a** and 1 h for **5b**)

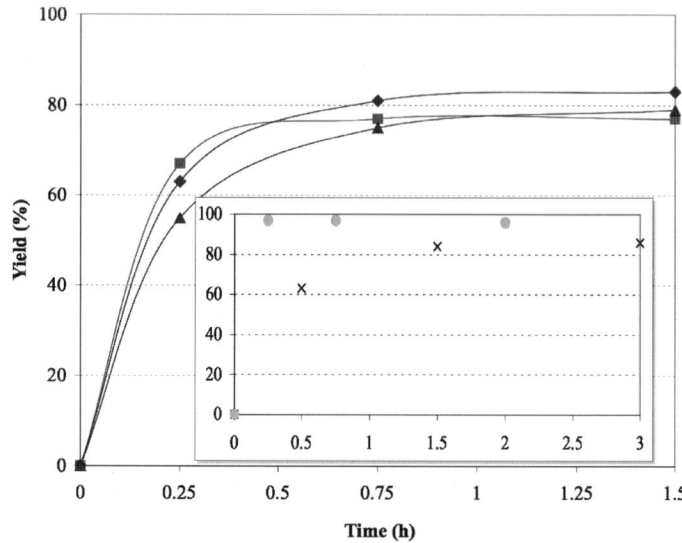

Fig. 4 Yield as a function of time for entry 1 in Fig. 3 (■ cycle 1, ♦ cycle 2, ▲ cycle 3), and (*inset*) for entries 3 (×) and 7 (●) in Fig. 3 (first cycle only)

could be run *in the absence* of solvent. The catalyst **5a** similarly precipitated from the product mixture, and yields for a four-cycle sequence were all > 95%.

In order to better gauge the extent of catalyst recovery [30, 31], the rate of the reaction of **2a** and **3** in octane was monitored as a function of cycle. For the first cycle, the maximum yield was reached after ca. 1 h, as illustrated in Fig. 4. The second and third cycles were similar. More importantly, when the yields are compared at partial conversion (0.25 h), only modest decreases are noted (67, 63, and 55%). Thus, ca. 90% of the activity is maintained from cycle to cycle.

The more basic and nucleophilic fluorous phosphine **5b** was similarly studied, and gave even better results. As summarized in Fig. 3 (entries 6 and 7), high yields of **4a,c** were obtained with much shorter run times (1 h vs 8 h). In the case of **4c**, the data were much improved over those with catalyst **5a** (entries 7 vs 3). The relative rates of these two reactions are compared in the inset in Fig. 4 (**5b** ≫ **5a**).

4
First Test System with Fluorous Support: Leaching Data [33,34]

The above procedures require the separation of a solid catalyst from a liquid phase or product. A potential problem, intrinsic to such methodologies, is that the amount of catalyst manipulated can be very small. For example, some catalyst precursors for the Heck reaction can be used at 0.0001 mol % levels, or even lower [38, 39]. On typical laboratory scales, these quantities can barely be visualized.

One way to increase the mass of recycled material is to use a support, as illustrated in sequence B of Fig. 1. At one extreme, the support could be inert, with only a mechanical function. At another level, the support might provide physical adhesion—for example, for a waxy or gumlike catalyst. At yet another level, the support might provide attractive interactions that would enhance recovery. Although such interactions between saturated fluorocarbons are very small, we sought to bring them into play. Hence, Teflon® shavings with average $x/y/z$ dimensions of 1–2 mm (Fig. 5) were selected for initial study.

Selected experiments from Fig. 3 were repeated, but in the presence of such Teflon® shavings. Yield data were comparable. Importantly, the recovered catalyst residues became more compact and firmer, despite the added mass of the Teflon®, suggestive of genuine physical adhesion or attraction. The catalyst residues were easier to manipulate, and less leaching occurred.

Leaching can be analyzed with respect to both the catalyst (rest state) and catalyst degradation products. The above reactions involve the separation of an n-octane product solution from a **5a** catalyst residue at – 30 °C, and a subsequent n-octane extraction at – 30 °C. The data in Fig. 2, together with the solvent quantities employed, predict catalyst leaching of < 0.33% per cycle (calculated from the solubility at – 20 °C). This rises to 1.0 and 3.6% if phase separations are conducted at 0 and 20 °C, respectively.

Leaching was first assayed for reactions conducted in the presence of Teflon® shavings. The n-octane supernatant from the reaction of **2a** and **3** (first cycle) was treated with internal standards for ^{31}P and ^{19}F NMR analyses. A ^{31}P NMR spectrum showed a barely integratable signal for the oxide of **5a** (41.9 ppm) [36], corresponding to ca. 0.2% leaching. No other signals could be detected. A ^{19}F NMR spectrum showed a much clearer CF$_3$ signal

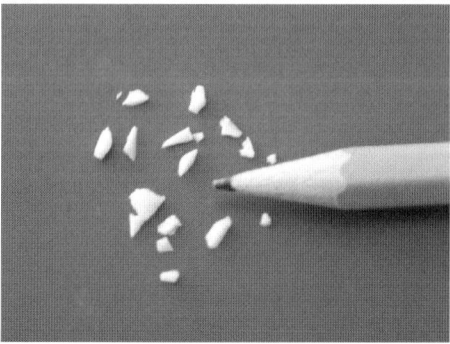

Fig. 5 Teflon® shavings used as support for the recycling of **5a**

amounting to 2.3% of the **5a** loading (normalized to the three ponytails). Note that all compounds of the formula $X(CF_2)_7CF_3$ give the same CF_3 NMR chemical shift. Hence, the "total number of leached ponytails"—which will include all catalyst degradation products—can be quite sensitively determined by this method.

The **5a**/Teflon® residue from one of the runs was treated with a phosphorus-containing internal standard. A ^{31}P NMR spectrum showed 85.2% of the original **5a** (– 24.5 ppm) and 12.2% of new phosphorus-containing species (97.4% phosphorus recovery). Some of these peaks may represent alternative rest states, and additional experiments on this point are detailed elsewhere [34]. An analogous experiment was conducted with fluorine-containing internal standard. A ^{19}F NMR spectrum indicated a 97.9% recovery of ponytails. Related experiments were conducted for identical reactions in the absence of Teflon® shavings. Higher leaching levels were always observed [34].

As an interim summary, the experiments in this section and the previous one were designed not so much with the aim of finding the optimum rate and yield conditions for catalysis. Rather, a paradigm by which coworkers could rationally evaluate and compare different catalysts and recovery protocols was sought. Future efforts should clearly focus on the more basic and nucleophilic catalyst **5b**, and/or analogs with a greater number of methylene groups (m) or longer R_{fn} segments. In addition, there are many additional possibilities for fluorous supports, as exemplified below.

5
Second Test System: Thermomorphic Metallacycle Catalysts [38–40]

Our attention was next drawn to palladium-catalyzed Heck condensations of aryl halides [38, 39]. These are commonly conducted at temperatures of 80–140 °C in higher-boiling solvents such as DMF. Under appropriate conditions all products and by-products remain soluble at room temperature. They therefore represent excellent test reactions for the protocols in Fig. 1. For homogeneous molecular catalysts, palladium halides are believed to be

7, X = **a**, OAc; **b**, Cl; **c**, I   **8**

the rest states. Many types of palladacycles have been successfully employed as catalyst precursors.

As detailed elsewhere, the fluorous palladacycle acetates and halides **7** and **8** were synthesized [38, 39]. These feature three R_{f8} ponytails, and were poorly soluble in common organic solvents at room temperature, and insoluble in DMF. However, they were very soluble in DMF at higher temperatures. All were effective catalyst precursors for Heck reactions (100–140 °C), and precipitated (as the halides) upon cooling. However, a number of control experiments established that **7** and **8** served as steady-state sources of colloidal palladium nanoparticles, formed anew with each cycle until the palladacycles were exhausted. These, or low-valent Pd(0) species derived therefrom, were the true catalysts.

We therefore turned to fluorous phosphorus/carbon/phosphorus (PCP) [40] and sulfur/carbon/sulfur (SCS) [41] pincer complexes. During the execution of this project, studies appeared [42, 43] strongly suggesting that most palladium pincer complexes employed as catalyst precursors for the Heck reaction first decompose to palladium nanoparticles or low-valent Pd(0) species. At the time, it was thought that the tridentate nature of these fluorous ligands might inhibit the formation of nanoparticles. Unfortunately, our first-generation PCP ligand proved much more difficult to synthesize than expected. As shown in Scheme 2, two usually reliable methods were troublesome. Free radical additions of primary phosphines to fluorous alkenes

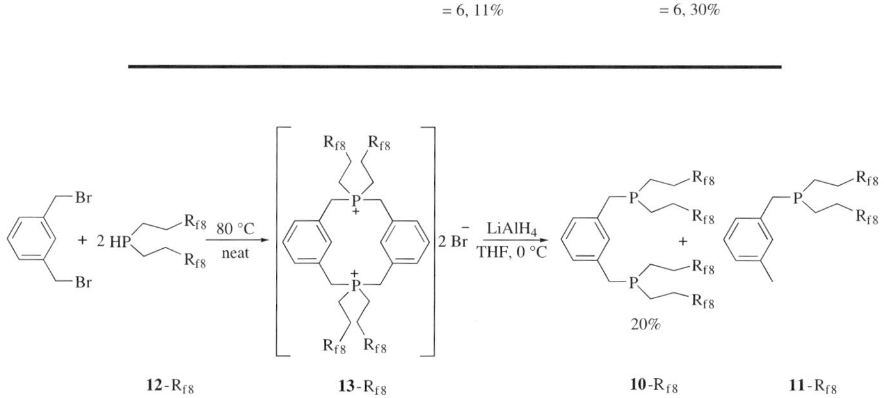

Scheme 2 Syntheses of fluorous PCP pincer ligands **10-R_{fn}**

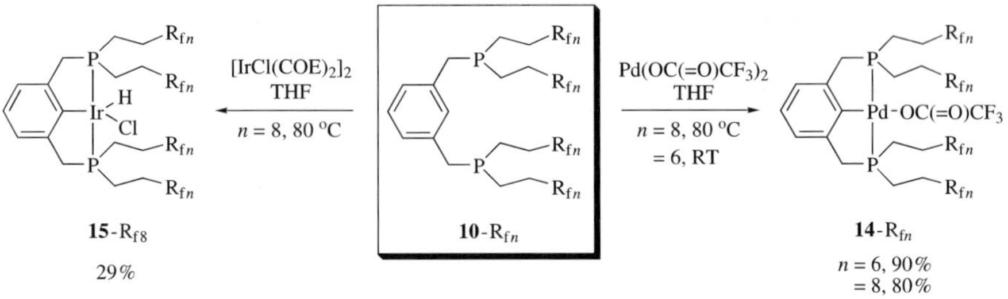

$[IrCl(COE)_2]_2$
THF
$n = 8, 80\,°C$

$Pd(OC(=O)CF_3)_2$
THF
$n = 8, 80\,°C$
$= 6, RT$

15-R_{f8}

29%

10-R_{fn}

14-R_{fn}

$n = 6, 90\%$
$= 8, 80\%$

Scheme 3 Syntheses of metal complexes of fluorous PCP pincer ligands **10**-R_{fn}

$H_2C = CHR_{fn}$ normally give tertiary phosphines in high yields [36]. However, in the case of the diprimary dibenzylic diphosphine **9** (Scheme 2, top), the major product was not the target PCP ligand **10**-R_{fn} but rather a tertiary *mono*phosphine (**11**-R_{fn})!

Similarly, secondary phosphines and α,α′-dibromo-*m*-xylene usually react to give diprotonated ditertiary diphosphines [44–48]. However, in the case of the fluorous secondary phosphine **12**-R_{f8} (Scheme 2, bottom), dialkylation occurred to give the metacyclophane **13**-R_{f8}. As detailed elsewhere [40], extensive efforts to adjust the stoichiometry to favor noncyclized products failed. Fortunately, the reduction of **13**-R_{f8} with LiAlH$_4$ gave some of the target ligand **10**-R_{f8}.

As shown in Scheme 3, the PCP ligand could be palladated in high yield. The resulting complexes **14**-R_{fn} were very soluble in fluorous solvents, insoluble in hexane, slightly soluble in ether, and moderately soluble in THF, CH$_2$Cl$_2$, and acetone. However, **14**-R_{fn} readily dissolved in hot hexane, and with **14**-R_{f8} single crystals were obtained upon cooling. X-ray analysis afforded the structures shown in Fig. 6.

In crystalline **14**-R_{f8}, one ponytail on each phosphorus atom extends above the palladium square plane, and the other below. Both run parallel to the ponytails on the *trans* phosphorus atom, which are nearly in van der Waals contact (Fig. 6, middle). In accord with the tendency for fluorous phases to segregate, the lattice is divided into fluorous and non-fluorous domains (Fig. 6, bottom), with the ponytails of neighboring molecules in comparable van der Waals contact.

For many processes, the solubility of **14**-R_{f8} is likely somewhat too high for efficient recycling as in Fig. 1. However, before undertaking the synthesis of **14**-R_{f10}, or other more highly fluorous derivatives, some Heck reactions were screened using **14**-R_{f8} as the catalyst precursor. The recyclability under fluorous/organic liquid/liquid biphase workup conditions was poor, and several observations indicated the formation of nanoparticles, in line with other recent observations [42, 43]. Analogous results were obtained in more extensive studies with SCS ligands, and these will be detailed separately.

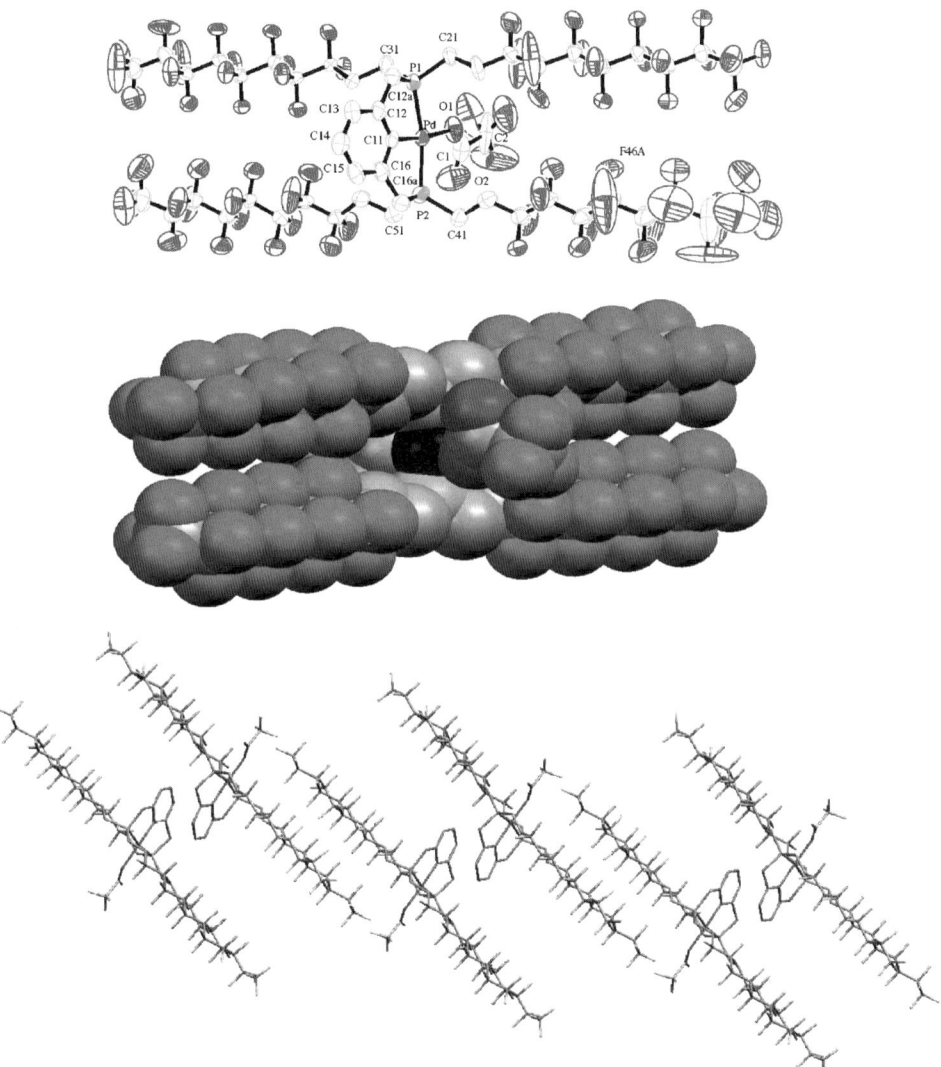

Fig. 6 Structure of the fluorous palladium PCP pincer complex **10-R$_{f8}$**: *top*, ORTEP representation; *middle*, view with atoms at van der Waals radii; *bottom*, packing diagram

Pincer complexes catalyze a variety of other organic reactions [49–51]. Hence, this work is currently being extended to other metals, and other more readily accessible PCP systems. For example, as shown in Scheme 3, **10-R$_{f8}$** can be converted to the iridium hydride chloride complex **15-R$_{f8}$**. Closely related dihydride complexes catalyze dehydrogenations of alkanes at high temperatures [52]. However, no efforts to develop recoverable catalysts have been reported to date.

6
Third Test System: Thermomorphic Rhodium Catalysts [53,54]

We sought to expand the range of catalysts and fluorous supports that could be applied in Fig. 1, and developed a surprisingly effective procedure involving common commercial Teflon® tape. As described below, this provides not only a convenient means of catalyst recovery, but also of catalyst delivery. From an engineering standpoint, tapes offer a variety of unique attributes, and it is reasonable to extrapolate that meshes and/or reactor parts such as liners could be fabricated with similar properties.

The ketone hydrosilylation shown in Fig. 7 was used as a test reaction. This can be catalyzed by the fluorous rhodium complexes 16-R_{f6} and 16-R_{f8} under fluorous/organic liquid/liquid biphase conditions [55, 56]. These red-orange compounds have very little or no solubility in organic solvents at room temperature [57]. However, their solubilities increase markedly with temperature. Several features render this catalyst system a particularly challenging test for recovery via precipitation. First, a variety of rest states are possible (e.g., various $Rh(H)(SiR_3)$ or $Rh(OR')(SiR_3)$ species), each with unique solubility properties. Second, the first cycle exhibits an induction period, indicating some fundamental alteration of the catalyst precursor.

Dibutyl ether was selected as the solvent due to its extended liquid range (bp 142 °C). A solution of cyclohexanone (**17**; 0.53 M) and $PhMe_2SiH$ (1.2 equiv) was treated with 1.0 mol% of **16**-R_{f6} or **16**-R_{f8} and warmed to 65 °C, achieving homogeneous conditions. After 8 h, the mixture was cooled

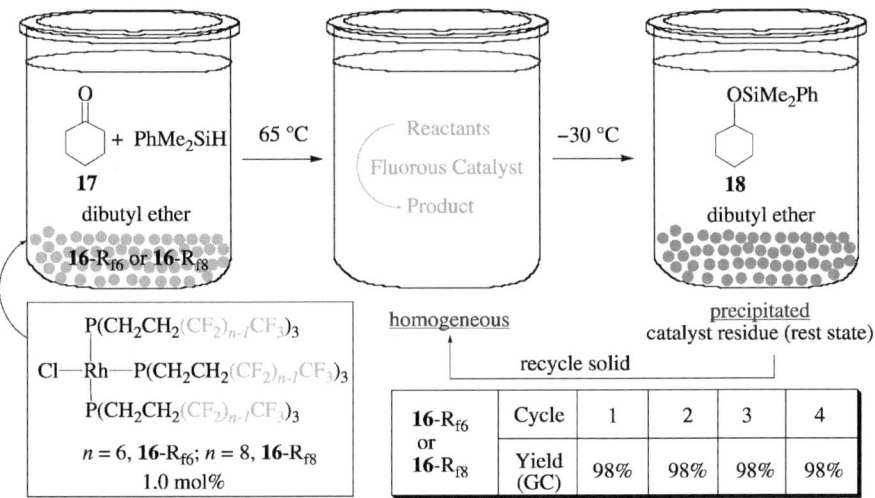

Fig. 7 Recycling of thermomorphic fluorous rhodium catalysts **16**-R_{fn} via solid/liquid phase separation

to room temperature. A brown-red catalyst residue slowly precipitated, and no color remained in the product-containing supernatant. Following workup (– 30 °C), GC analysis indicated a 98% yield of the silyl ether **18**. The residue was charged with fresh reactants and the cycle repeated three additional times, each giving a 98% yield of **18**.

This hydrosilylation can be run at much lower catalyst loadings than the phosphine-catalyzed addition in Scheme 2 and Fig. 3. Hence, supports were sought to facilitate recovery. A similar sequence was conducted with 0.15 mol% of **16**-R_{f6}, but in the presence of Teflon® tape (thickness/width 0.0075/12 mm). Homogeneous conditions could be achieved at 55 °C, and photographs of a typical sequence are shown in Fig. 8. The white tape became lightly colored during the reaction, and orange-red when the sample was cooled. Enthalpic interactions between perfluoroalkyl segments are very small. Hence, it is noteworthy that the catalyst rest state phase separates *onto* the tape, as opposed to giving a second solid phase. Interestingly, the Teflon® stirring bar remained white, perhaps due to a processing step or coating.

Analysis as above indicated a 97% yield of **18**. An identical reaction was conducted, and the concentrations of **17** and **18** were monitored over four cycles, as summarized in Fig. 9. As in earlier work [55, 56], the first cycle ex-

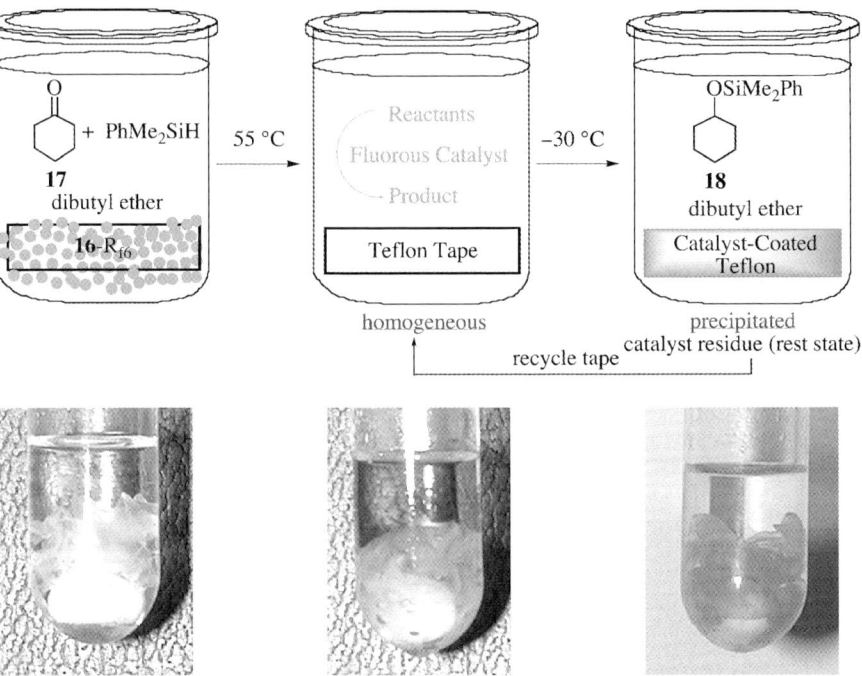

Fig. 8 Recycling of the thermomorphic fluorous rhodium catalyst **16**-R_{fn} using Teflon® tape

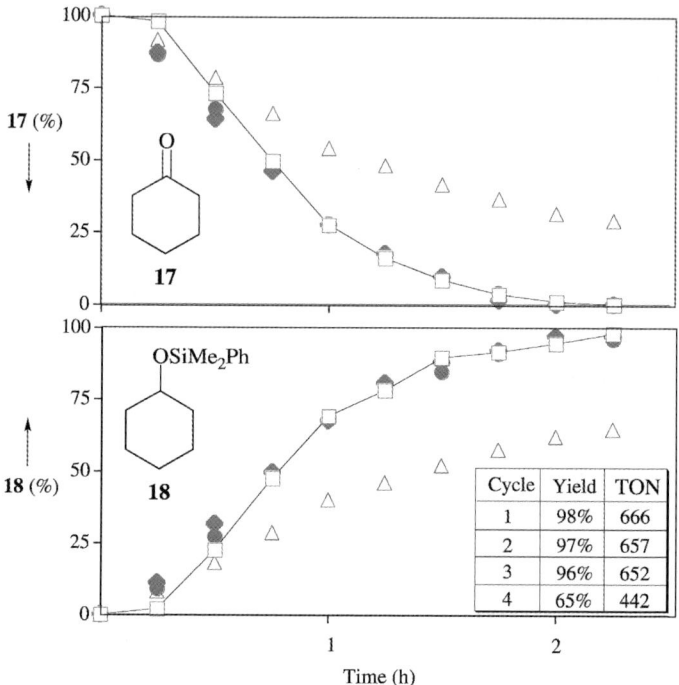

Fig. 9 Reaction profile for four cycles under the conditions of Fig. 8: □ cycle 1; ◆ cycle 2; ● cycle 3; △ cycle 4. TON: turnover number

hibited an induction period. Retention of activity was excellent in the second and third cycles, but there was substantial loss in the fourth. In view of the similar rates of the first three cycles, this can be attributed to catalyst deactivation. An intrinsic problem with the recycling protocol, such as leaching, should afford a relatively constant loss of activity per cycle.

Leaching was probed in two ways. First, the dibutyl ether supernatants from the first three cycles in Fig. 9 were combined, and the total ponytail leaching assayed by ^{19}F NMR as above. A signal corresponding to 11.4% of the phosphine ligands of **16-R$_{f6}$** was observed. Since the active and resting states of the catalyst likely involve two phosphine ligands, some extraction is not surprising. Second, the supernatants were analyzed for rhodium. The first cycle showed leaching corresponding to 0.57% of the original charge, and the second cycle 5.3%. We attribute the increased value to the onset of catalyst degradation.

One obvious procedural refinement would be to precoat the catalyst on the Teflon® tape. This would allow low loadings to be delivered by *length* as opposed to mass measurements, or the tedious preparation of standard solutions. Accordingly, strips of tape were added to a solution of **16-R$_{f6}$** in $CF_3C_6F_{11}$. The solvent was removed under an inert gas stream to give a yellowish catalyst-coated tape. This could be applied in a three-cycle sequence with yield results similar to those in Fig. 9.

When catalysts are recycled as solid residues, it is important to exclude impurities that may "piggyback"—such as metal particles—as the active species. This was probed in two ways. First, the tape was removed after a first cycle, rinsed, and transferred to a new vessel. A second charge of **17** and dibutyl ether was added, but *not* the PhMe$_2$SiH. The sample was warmed to 55 °C, the now off-white tape was "fished out", and PhMe$_2$SiH was added. The rate profile was similar to the first cycle (ca. 20% slower at higher conversions), consistent with predominant homogeneous catalysis by desorbed fluorous species. Second, the second cycle of a sequence was conducted in the presence of elemental mercury, which inhibits catalysis by metal particles [57]. However, the rate profile was the same as a sequence in the absence of mercury.

Similar results have been obtained with other ketones such as 2-octanone, acetophenone, and benzophenone. The basis for catalyst deactivation remains under study, together with alternative fluoropolymer supports [59]. Analogous adsorption phenomena have been found with other fluorous metal complexes, as will be detailed in future reports. In our opinion, there are many additional possible applications for this catalyst recovery/delivery technology, and these are under active pursuit.

7
Results from Other Investigators

7.1
Recycling Without Supports

Other research groups have also made significant contributions to the themes described above. Fluorous catalysts that have been recovered from homogeneous reactions in organic solvents at elevated temperatures by simple precipitations (sequence A-I, Fig. 1) are summarized in Fig. 10 (**20–25**). A representative application for each is given in Scheme 4.

To our knowledge, the first such catalyst described was perfluorinated di(*n*-octyl)ketone (**20**) [5]. Sheldon found that this compound dissolved in refluxing 10 : 1 v/v ClCH$_2$CH$_2$Cl/EtOAc, under which conditions it catalyzed the epoxidation of alkenes by H$_2$O$_2$ (reaction A, Scheme 4). Upon cooling to 0 °C, **20** was recovered in 92% yield. We overlooked this work in our earliest papers, together with an isolated example by Curran [6]. As is often the case in science, the potential generality of a phenomenon was not recognized in the initial reports.

Concurrently with our first communication, Ishihara and Yamamoto reported that the fluorous phenylboronic acid **21** efficiently catalyzed con-

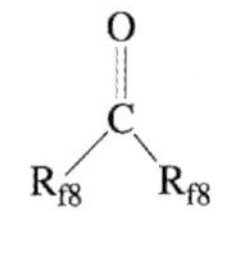

$$\underset{20}{\overset{O}{\underset{R_{f8}}{\|}}\underset{R_{f8}}{C}}$$

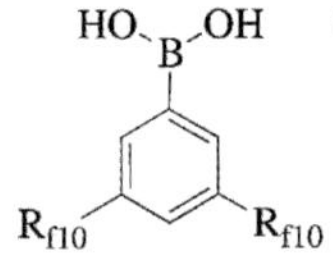

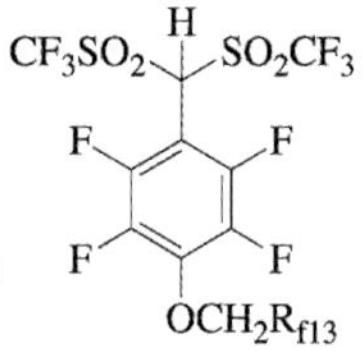

$$\underset{21}{\overset{HO}{\underset{R_{f10}}{}}\overset{OH}{\underset{R_{f10}}{}}}$$

$$\underset{22}{\overset{H}{\underset{OCH_2R_{f13}}{}}}$$

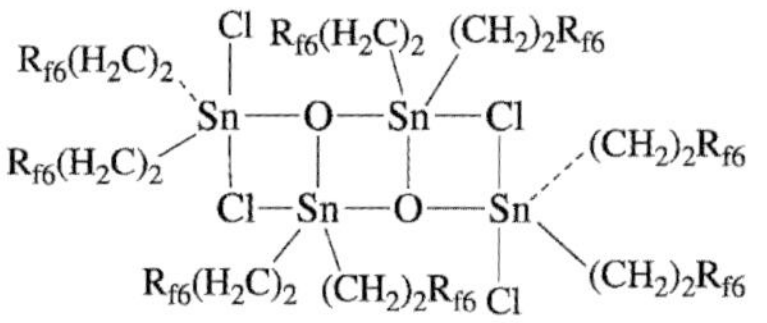

23

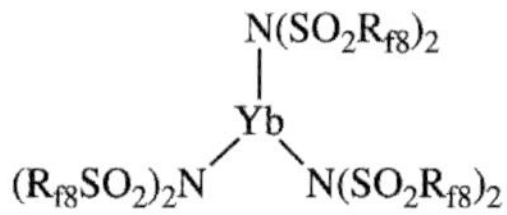

24

$$25 = 12[(R_{f8}(CH_2)_3)_3NCH_3]^+ [WZn_3(H_2O)_2(ZnW_9O_{34})_2]^{12-}$$

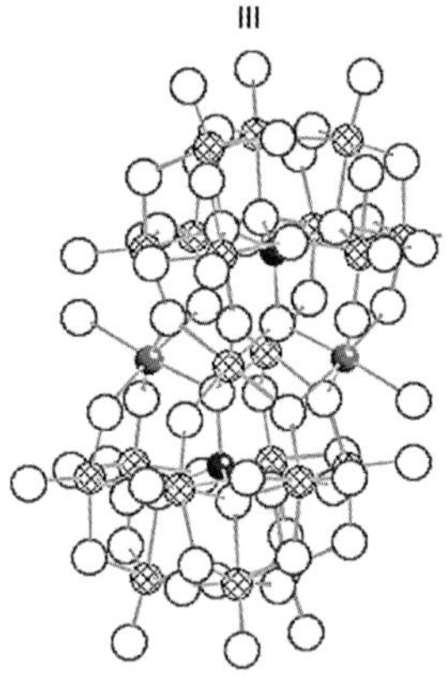

Fig. 10 Additional thermomorphic fluorous catalysts that have been recovered by solid/liquid phase separations

densations of carboxylic acids and amines to give amides (reaction B, Scheme 4) [7]. Reactions could be effected under homogeneous conditions in refluxing toluene or xylene, and **21** could be recovered by precipitation at room temperature. Ten such cycles were conducted, giving the amide in 96% isolated yield (> 99% conversion per cycle) and **21** in 26% yield (88% recovery per cycle). More recently, this group found that the fluorous bis(sulfone) **22**, which is a super Brønsted acid, catalyzes acetal formation (reaction C, Scheme 4) [10]. Reactions were effected under homogeneous conditions in refluxing hexane, and **22** could be recovered in 96% yield at room temperature. Benzoylations of alcohols and esterifications of carboxylic acids were similarly conducted in toluene and methanol, with 70–68% recoveries of **22**.

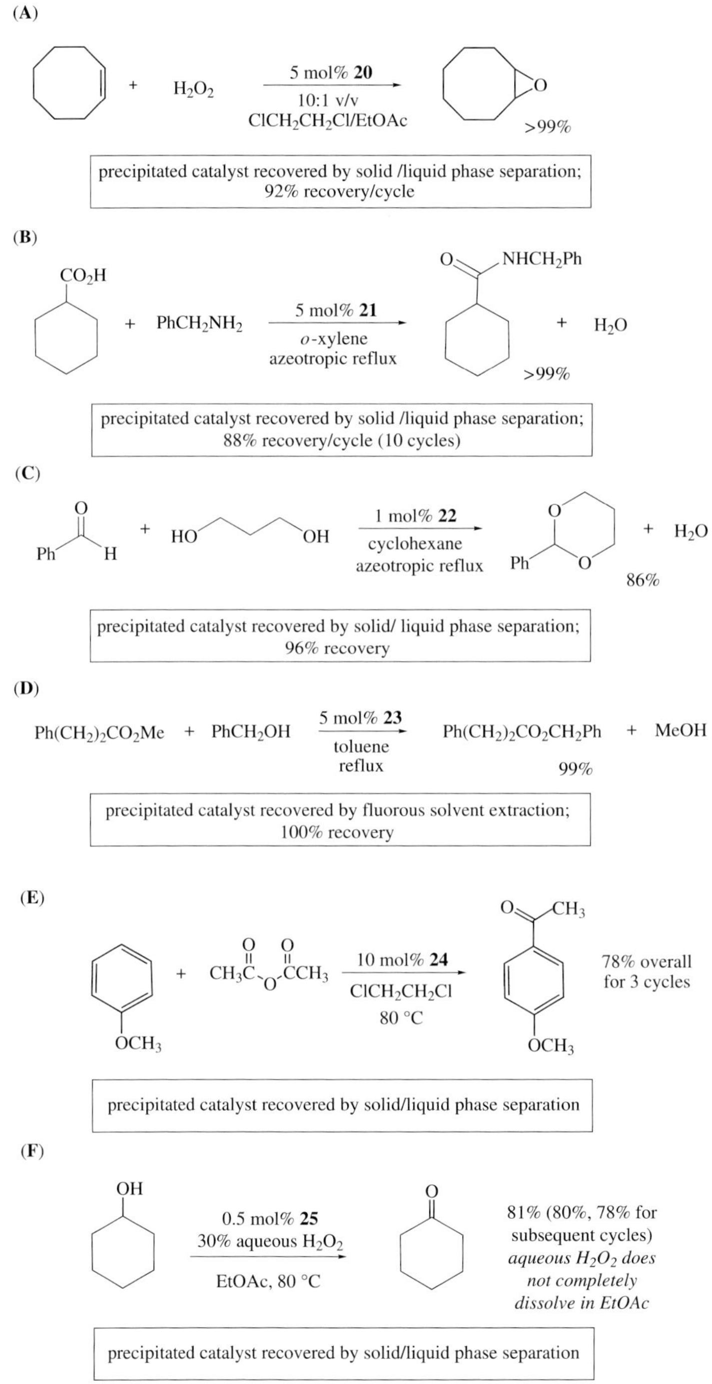
(A)
+ H2O2
5 mol% 20
10:1 v/v
ClCH2CH2Cl/EtOAc
O
>99%
precipitated catalyst recovered by solid /liquid phase separation;
92% recovery/cycle
(B)
CO2H
+ PhCH2NH2
5 mol% 21
o-xylene
azeotropic reflux
O
NHCH2Ph
+ H2O
>99%
precipitated catalyst recovered by solid /liquid phase separation;
88% recovery/cycle (10 cycles)
(C)
O
Ph
H
+ HO
OH
1 mol% 22
cyclohexane
azeotropic reflux
O
Ph
O
+ H2O
86%
precipitated catalyst recovered by solid/ liquid phase separation;
96% recovery
(D)
Ph(CH2)2CO2Me + PhCH2OH
5 mol% 23
toluene
reflux
Ph(CH2)2CO2CH2Ph + MeOH
99%
precipitated catalyst recovered by fluorous solvent extraction;
100% recovery
(E)
OCH3
+
O O
CH3C O CCH3
10 mol% 24
ClCH2CH2Cl
80 °C
O
CH3
OCH3
78% overall
for 3 cycles
precipitated catalyst recovered by solid/liquid phase separation
(F)
OH
0.5 mol% 25
30% aqueous H2O2
EtOAc, 80 °C
O
81% (80%, 78% for
subsequent cycles)
aqueous H2O2 does
not completely
dissolve in EtOAc
precipitated catalyst recovered by solid/liquid phase separation

Scheme 4 Representative applications of the catalysts in Fig. 10

Otera has reported that fluorous distannoxanes such as **23**, which dissociate to give Lewis acidic species, catalyze transesterifications in organic/fluorous solvent mixtures [8, 9]. Although **23** was insoluble in toluene at room temperature, it dissolved at reflux and efficiently promoted the transformation in reaction D of Scheme 4, as well as others. The catalyst precipitated upon cooling, but a fluorous solvent extraction was utilized for recovery (100%). Another thermomorphic fluorous Lewis acid catalyst was developed by Mikami [11]. He found that the ytterbium tris(sulfonamide) **24** could be used for Friedel–Crafts acylations under homogeneous conditions in $ClCH_2CH_2Cl$ at 80 °C, and precipitated upon cooling to −20 °C (reaction E, Scheme 4).

Neumann and Fish have studied the novel polyoxometalate salt **25**, which features 12 fluorous ammonium cations [12]. This material was insoluble in EtOAc (and toluene) at room temperature, but dissolved at 80 °C to give an effective catalyst system for the oxidation of alkenes and alcohols by 30% aqueous H_2O_2. Cooling precipitated the catalyst, which was reused. Additional examples of thermomorphic fluorous catalysts have been briefly described in meeting abstracts [60, 61] and will likely soon appear in the peer-reviewed literature.

7.2
Recycling With Supports

As would be expected, fluorous compounds are preferentially retained on fluorous silica gel [62]. Similarly, fluorous catalysts can be adsorbed on fluorous silica gel. These materials have been applied to reactions in organic solvents and water, both at room temperature and above [63–69]. The investigators have usually interpreted the transformations as "bonded fluorous phase catalysis", which corresponds to sequence B-II in Fig. 1. However, there remains the possibility that at least some catalysis proceeds under homogeneous conditions via desorbed species. To our knowledge, "fish-out" experiments analogous to that conducted with the Teflon® tape in Fig. 8 have not been conducted.

To date, reports have involved palladium catalysts for Suzuki and Sonogashira coupling reactions [63–66], rhodium catalysts for silylations of alcohols by trialkylsilanes [67, 68], and tin-, hafnium-, and scandium-based Lewis acid catalysts for Baeyer–Villiger and Diels–Alder reactions [69]. Regardless of exact mechanism, this recovery strategy represents an important direction for future research and applications development. Finally, a particularly elegant protocol where CO_2 pressure is used instead of temperature to desorb a fluorous rhodium hydrogenation catalyst from fluorous silica gel deserves emphasis [28].

8
Summary

The preceding examples clearly demonstrate that fluorous catalysts can be applied in single-phase, homogeneous reactions at elevated temperatures in the absence of fluorous solvents, and efficiently recovered via simple solid/liquid phase separations at lower temperatures. This procedure significantly expands the scope of fluorous biphase catalysis as originally formulated [1], and is distinctly "greener". Although organic solvents are commonly employed, there are also several cases where reactions and catalyst recovery can be effected under solvent-free conditions.

Some potential limitations associated with this protocol merit note. For example, with sequence A in Fig. 1, insoluble by-products will interfere with catalyst recovery. With sequence B, interference will depend upon the type of support. For instance, the Teflon® tape in Fig. 8 should be easily separable from another solid material, as would a mesh or reactor liner. Also, since heating is required to achieve homogeneity, the method is best suited for reactions conducted at elevated temperatures. However, there are many reactions which proceed rapidly under fluorous/organic liquid/liquid biphase conditions (i.e., before the miscibility temperature is reached) [55–57, 70]. Therefore, it is not unreasonable to expect that solid fluorous catalysts with little or no solubility can also efficiently promote certain reactions, as represented by sequence A-I in Fig. 1 [29].

To our knowledge, there have been no previous attempts to develop a broad class of *molecular* catalysts that have temperature-dependent solubilities. When molecular catalysts are covalently bound to polymeric supports, they generally assume the solubility properties of the host polymer. In the above fluorous catalysts, we like to think that a short segment of polymer is being grafted onto a molecular catalyst. In other words, the ponytails can be viewed as pieces of Teflon®, which impart more and more of the solubility characteristics of the polymer as they are lengthened.

In conclusion, the new recycling procedures described above offer virtually unlimited possibilities for optimization and "catalyst engineering". The lengths and other structural features of the ponytails are easily varied. There are innumerable types of possible fluoropolymer supports, as well many additional classes of fluorous supports. Accordingly, a variety of further refinements and developments can be expected in the near future.

Acknowledgements We thank the Bundesministerium für Bildung und Forschung (BMBF) for support.

References

1. Horváth IT, Rábai J (1994) Science 266:72
2. Gladysz JA (2004) Overview of structural classes of ponytails. In: Gladysz JA, Curran DP, Horváth IT (eds) Handbook of fluorous chemistry. Wiley, Weinheim, Chap 5
3. Hughes RP, Trujillo HA (1996) Organometallics 15:286
4. Gladysz JA, Emnet C (2004) In: Gladysz JA, Curran DP, Horváth IT (eds) Handbook of fluorous chemistry. Wiley, Weinheim, pp 11–23
5. van Vliet MCA, Arends IWCE, Sheldon RA (1999) Chem Commun 263
6. Olofsson K, Kim SY, Larhed M, Curran DP, Hallberg A (1999) J Org Chem 64:4539 (see Table 2, entry 4)
7. Ishihara K, Kondo S, Yamamoto H (2001) Synlett 1371
8. Xiang J, Orita A, Otera J (2002) Adv Synth Catal 344:84
9. Otera J (2004) Acc Chem Res 37:288
10. Ishihara K, Hasegawa A, Yamamoto H (2002) Synlett 1299
11. Mikami K, Mikami Y, Matsuzawa H, Matsumoto Y, Nishikido J, Yamamoto F, Nakajima H (2002) Tetrahedron 58:4015
12. Maayan G, Fish RH, Neumann R (2003) Org Lett 5:3547
13. Bergbreiter DE, Chandran R (1987) J Am Chem Soc 109:174
14. Bergbreiter DE, Zhang L, Mariagnanam VM (1993) J Am Chem Soc 115:9295
15. Bergbreiter DE, Case BL, Liu YS, Caraway JW (1998) Macromolecules 31:6053
16. Breuzard JAJ, Tommasino ML, Bonnet MC, Lemaire M (2000) J Organomet Chem 616:37
17. Dickerson TJ, Reed NN, Janda KD (2002) Chem Rev 102:3325
18. Bergbreiter DE (2002) Chem Rev 102:3345
19. Bosanac T, Yang J, Wilcox CS (2001) Angew Chem Int Ed 40:1875
20. Bosanac T, Yang J, Wilcox CS (2001) Angew Chem 113:1927
21. Bosanac T, Wilcox CS (2001) Tetrahedron Lett 42:4309
22. Bosanac T, Wilcox CS (2001) J Chem Soc Chem Commun 1618
23. Bosanac T, Wilcox CS (2002) J Am Chem Soc 124:4194
24. Honigfort ME, Brittain WJ, Bosanac T, Wilcox CS (2002) Macromolecules 35:4849
25. Wilcox CS, Bosanac T (2004) Org Lett 6:2321
26. Koch D, Leitner W (1998) J Am Chem Soc 120:13398
27. Francio G, Wittman K, Leitner W (2001) J Organomet Chem 621:130
28. Ablan CD, Hallett JP, West KN, Jones RS, Eckert CA, Liotta CA, Jessop PG (2003) Chem Commun 2972
29. Peng Z, Orita A, An D, Otera J (2005) Tetrahedron Lett 46:3187
30. Gladysz JA (2001) Pure Appl Chem 73:1319
31. Gladysz JA (2002) Chem Rev 102:3215
32. Gladysz JA, da Costa RC (2004) In: Gladysz JA, Curran DP, Horváth IT (eds) Handbook of fluorous chemistry. Wiley, Weinheim, pp 24–40
33. Wende M, Meier R, Gladysz JA (2001) J Am Chem Soc 123:11490
34. Wende M, Gladysz JA (2003) J Am Chem Soc 125:5861
35. Inanaga J, Baba Y, Hanamoto T (1993) Chem Lett 241
36. Alvey LJ, Rutherford D, Juliette JJJ, Gladysz JA (1998) J Org Chem 63:6302
37. Alvey LJ, Meier R, Soós T, Bernatis P, Gladysz JA (2000) Eur J Inorg Chem 1975
38. Rocaboy C, Gladysz JA (2002) Org Lett 4:1993
39. Rocaboy C, Gladysz JA (2003) New J Chem 27:39
40. Tuba R, Tesevic V, Dinh LV, Hampel F, Gladysz JA (2005) Dalton Trans 2275
41. da Costa RC (2006) Doctorial Dissertation, Universität Erlangen-Nürnberg

42. Yu K, Sommer W, Richardson JM, Weck M, Jones CW (2005) Adv Synth Catal 347:161
43. Bergbreiter DE, Osburn PL, Frels J (2005) Adv Synth Catal 347:172
44. Gusev DG, Madott M, Dolgushin FM, Lyssenko KA, Antipin MY (2000) Organometallics 19:1734
45. Hollink E, Stewart JC, Wei P, Stephan DW (2003) J Chem Soc Dalton Trans 3968
46. Grimm JC, Nachtigal C, Mack HG, Kaska WC, Mayer HA (2000) Inorg Chem Commun 3:511
47. Mohammad HAY, Grimm JC, Eichele K, Mack HG, Speiser B, Novak F, Quintanilla MG, Kaska WC, Mayer HA (2002) Organometallics 21:5775
48. Hermann D, Gandelman M, Rozenberg H, Shimon LJW, Milstein D (2002) Organometallics 21:812
49. Albrecht M, van Koten G (2001) Angew Chem Int Ed 40:3750
50. Albrecht M, van Koten G (2001) Angew Chem 113:3866
51. van der Boom ME, Milstein D (2003) Chem Rev 103:1759
52. Jensen CM (1999) Chem Commun 2443
53. Dinh LV, Gladysz JA (2005) Angew Chem Int Ed 44:4095
54. Dinh LV, Gladysz JA (2005) Angew Chem 117:4164
55. Dinh LV, Gladysz JA (1999) Tetrahedron Lett 40:8995
56. Dinh LV, Gladysz JA (2005) New J Chem 29:173
57. Juliette JJJ, Rutherford D, Horváth IT, Gladysz JA (1999) J Am Chem Soc 121:2696
58. Widegren WA, Finke RG (2003) J Mol Catal A 198:317
59. Jurisch M (2005) Diploma thesis, Universität Erlangen-Nürnberg
60. Contel M, Villuendas PR, Fernández-Gallardo J, Alonso PJ, Vincent JM, Fish RH (2005) Inorg Chem 44:9771
61. Lantos D, Contel M, Sanz S, Horváth IT (2005) In: Book of abstracts, 1st international symposium on fluorous technologies, Bordeaux, 3–6 July 2005
62. Curran DP (2004) In: Gladysz JA, Curran DP, Horváth IT (eds) Handbook of fluorous chemistry. Wiley, Weinheim, Chap 7
63. Tzschucke CC, Markert C, Glatz H, Bannwarth W (2002) Angew Chem Int Ed 41:4500
64. Tzschucke CC, Markert C, Glatz H, Bannwarth W (2002) Angew Chem 114:4678
65. Tzschucke CC, Bannwarth W (2004) Helv Chim Acta 2882
66. Tzschucke CC, Andrushko V, Bannwarth W (2005) Eur J Org Chem :5248
67. Biffis A, Zecca M, Basato M (2003) Green Chem 5:170
68. Biffis A, Braga M, Basato M (2004) Adv Synth Catal 346:451
69. Yamazaki O, Hao X, Yoshida A, Nishikido J (2003) Tetrahedron Lett 44:8791
70. Rutherford D, Juliette JJJ, Rocaboy C, Horváth IT, Gladysz JA (1998) Catal Today 42:381

Top Organomet Chem (2008) 23: 91–108
DOI 10.1007/3418_2008_069
© Springer-Verlag Berlin Heidelberg
Published online: 26 April 2008

Regulated Systems for Catalyst Immobilisation Based on Supercritical Carbon Dioxide

Jens Langanke · Walter Leitner (✉)

Institut für Technische und Makromolekulare Chemie, RWTH Aachen University,
Worringerweg 1, 52074 Aachen, Germany
leitner@itmc.rwth-aachen.de

Abstract Promising strategies to use supercritical CO$_2$ (scCO$_2$) for catalyst immobilisation can be developed on the basis of its unique and tunable solubility properties in the vicinity of, and especially just beyond, the critical data of pure carbon dioxide. The present chapter describes briefly the background and experimental details of four different methods. In the first case, scCO$_2$ is used as a "switch" to precipitate homogeneous catalysts and extract the organic components. Biphasic systems, where an immiscible liquid provides a permanent stationary phase for the catalyst, are exemplified for polyethylene glycol (PEG) and ionic liquids as the catalyst phase. Finally, so-called inverted biphasic systems, where scCO$_2$ serves as the stationary phase for the catalyst, are presented with water as the mobile substrate/product phase.

Keywords Biphasic systems · Catalyst immobilisation · Ionic liquids · Multiphase catalysis · Supercritical carbon dioxide

1
Introduction

The immobilisation of homogeneous catalysts is an intensely investigated research area in academia and industry aimed at finding novel and sustainable solutions to the most fundamental problem of the art of homogeneous catalysis: the simple separation of products and catalyst and direct catalyst reuse with a minimum of additional—or even better—no further working steps [1, 2]. The use of compressed and supercritical CO_2 (scCO_2) is a promising approach to this problem, which is one of the major driving forces in the continuously growing interest in this field. A number of different strategies to separate homogeneous catalysts from products in the presence of CO_2 during or after the reaction have emerged, and in this section some promising examples will be discussed. They have been selected to demonstrate how the interplay between the molecular structure of the catalyst and the unique physico-chemical properties of scCO_2 can be exploited to regulate and control the solubility and partitioning of catalysts and substrates/products in multiphasic systems.

The basic physico-chemical properties of scCO_2 for application in organometallic catalysis have been the subject of detailed reviews in recent years [3–7]. In addition to the environmental and safety benefits of using non-toxic, non-flammable and innocuous carbon dioxide as solvent, several other aspects emerge from these studies. For example, the opportunity to operate the reaction step under truly homogeneous conditions can be an incentive for using scCO_2 as a reaction medium in molecular catalysis. This is especially attractive for reactions in which one reactant is a compressed gas. Hydrogenation, hydroformylation or oxidations are examples which in conventional reaction media may suffer from mass transfer limitations owing to restricted gas solubility in the liquid phase. As scCO_2 is completely miscible over all composition ranges with the majority of reactive gases used in chemical synthesis, the reaction can be operated in a truly homogeneous single-phase medium. Additionally, the density of the medium may be controlled through alteration of pressure or temperature, thus allowing very fine control of the solvent properties which can influence reactivities and selectivities in complex catalytic cycles. Finally, in certain cases CO_2 may interact chemically with the solutes; this effect may be used to aid the system, for example where it acts as a protecting group.

The use of scCO_2 for catalyst immobilisation is associated directly with its unique and tunable solubility properties in the vicinity of, and especially just beyond, the critical data of pure carbon dioxide ($T_C = 31.0\,°C$, $p_C = 73.75$ bar, $d_C = 0.467\,g \cdot mL^{-1}$). The solubility of a solute in scCO_2 is governed mainly by its vapour pressure (volatility) and its polarity. Typically, high volatility and low polarity result in high solubility. Thus, organometallic complexes and materials such as those used in homogeneous catalysis have generally very limited solubilities in scCO_2 as compared to many organic substrates and products. As will be exemplified in Sect. 2 of this chapter, this difference can be exploited for a separation strategy where scCO_2 is used as a solubility switch in homogeneous catalysis. The solubility difference can be enhanced further by introducing an immiscible liquid into the system that provides a permanent stationary phase for the catalyst (Sects. 3 and 4). However, the intrinsic solubility can be also "overruled" by the introduction of highly CO_2-philic [8] groups into the catalyst structure. This allows the implementation of so-called inverted biphasic systems, where scCO$_2$ serves as the stationary phase for the catalyst. In Sect. 5, the application of this strategy to the conversion of highly water-soluble substrates and products will be presented.

2
Catalyst Separation and Recycling Using scCO$_2$ as Solubility Switch

A straightforward approach to using CO$_2$ for separation in homogeneously catalysed processes can be applied when the catalyst is soluble in the reactant/product but insoluble in scCO$_2$. In these cases the reaction can be carried out neat in the organic phase in the absence of CO$_2$. After the reaction is completed, CO$_2$ is introduced under suitable conditions to precipitate the catalyst. The products are then extracted with scCO$_2$, while the CO$_2$-insoluble catalyst remains in the reactor and is immediately ready for reuse. Several variants to utilise scCO$_2$ to extract products for the recycling of poorly soluble catalysts have been described [9], including cases in which scCO$_2$ was also present during the reaction [10, 11]. In these cases, the conditions are set in such a way that the catalyst is at least partly soluble in scCO$_2$ during the reaction, but practically insoluble in the extraction step.

A very efficient example of the approach to introduce scCO$_2$ as a switch after the neat reaction employs catalysts with ligands carrying polyethylene glycol (PEG) side chains. The strategy is to use PEG-substituted phosphine ligands for the formation of organometallic complexes, which are completely insoluble in scCO$_2$ but soluble in a wide range of organic compounds involved in the reaction. With ligand MeO-PEG$_{750}$PPh$_2$ (1) [12] it was shown that the hydroformylation of 1-octene catalysed by a soluble catalyst formed in situ from 1 and [Rh(acac)(CO)$_2$] (acac = acetylacetonate; P : Rh = 5 : 1) could be "switched off" completely with the introduction of CO$_2$ into the reactor.

1

For example, conversion after two hours reaction time ($T = 70\,^\circ$C, $p(\text{H}_2/\text{CO}) = 50$ bar) dropped from 99% in the absence of CO$_2$ to 66.4% at a density $d(\text{CO}_2) = 0.35\,\text{g} \cdot \text{mL}^{-1}$ to 0% at $d(\text{CO}_2) = 0.57\,\text{g} \cdot \text{mL}^{-1}$. The observation of a yellow-orange solid precipitated during the addition of CO$_2$ confirmed the efficient separation of catalyst and substrate. The same separation was induced in the product mixture if CO$_2$ was introduced after the hydroformylation was completed. The mixture of regioisomeric nonanals formed during the reaction was extracted quantitatively with scCO$_2$ leaving an active hydroformylation catalyst behind in the reactor.

Consequently a "cartridge" system for the hydroformylation of a variety of different alkenes using a single batch of catalyst was developed to fully exploit the potential of this approach (Scheme 1). Four different alkenes were applied in total using the same batch of catalyst for nine consecutive catalytic cycles. Conversion was driven to completeness in all cases, with selectivities remaining unchanged. Only 1.2% of rhodium and 2.4% of phosphorus was lost in nine cycles. Representative results are shown in Table 1.

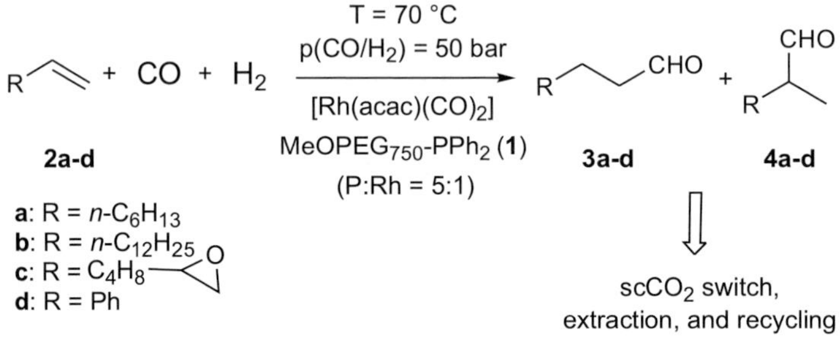

Scheme 1 Hydroformylation with 1/[Rh(acac)(CO)$_2$]/scCO$_2$

Furthermore, it was shown that even different reactions could be carried out using the same batch of catalyst for various transformations.

Table 1 Hydroformylation of structurally diverse olefins **2a–d** using the cartridge catalysis system MeOPEG$_{750}$-PPh$_2$ (**1**)/[Rh(acac)(CO)$_2$]/scCO$_2$

Cycle[a]	Substrate	Sub/Rh	Conversion [%]	Major product, selectivity [%]	scCO$_2$ extraction[b] T [°C]/p [bar]/V [L]	Recovery [%]
1	2a	984	> 99	3a, 73.7	45–50/80–96/130	87.0
2	2a	1010	> 99	3a, 72.2	45–50/85–95/140	92.7
3	2b	1018	> 99	3c, 70.6	45–50/120–130/188	99.8
4	2b	1003	> 99	3c, 70.6	45–50/120–140/200	100
5	2c	997	> 99	3d, 68.8	45–50/90–100/120	100
6	2c	1010	> 99	3d, 68.9	45–50/95–105/130	100
7	2d	1039	> 99	4e, 90.0	45–50/86–95/115	91.0
8	2d	1015	> 99	4e, 89.8	45–50/88–100/120	91.0
9	2a	1056	94.1	3a, 73.0	45–50/80–100/100	86.2

[a] All reactions were carried out in a window-equipped stainless steel reactor ($V = 24$ mL) with the same catalyst batch generated in situ in the first cycle from [Rh(acac)(CO)$_2$] (0.01 mmol) and ligand **1** (0.05 mmol); reaction conditions: $T = 70$ °C, p(CO/H$_2$) = 50 bar, $t = 2$ h
[b] The exit flow was maintained at 0.5–0.7 L/min; the total volume of CO$_2$ is given as litres of gas at standard conditions

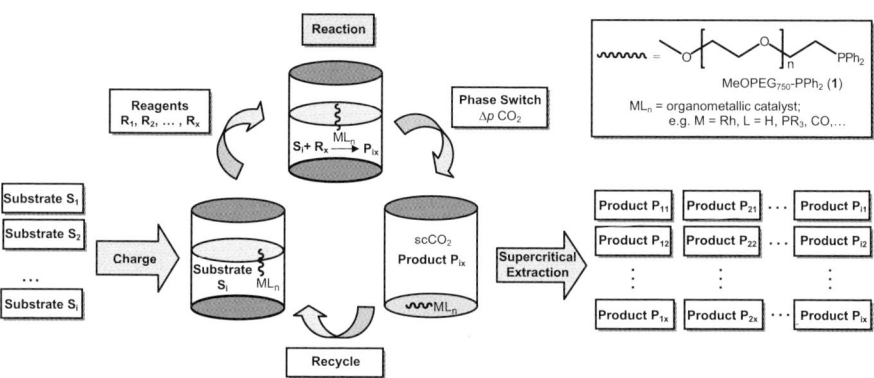

Fig. 1 Organometallic cartridge catalysis. A selection from a variety of transition metal complex catalysed conversions of substrates S$_i$ with different reagents R$_x$ to the desired product P$_{ix}$ is possible using a single catalyst batch with the same apparatus and separation techniques

As an example, styrene was subjected to hydroboration, hydroformylation, hydrogenation and finally hydroboration once more, using the same [Rh(acac)(CO)$_2$]/**1** catalyst system throughout the whole sequence. At the end of each reaction cycle the products were extracted using scCO$_2$ and the reactor was charged with fresh substrate. The conversions and selectivities obtained made clear that the catalyst is as active for the last hydroboration cycle as in the first run. Again, catalyst leaching in the products was very low and no significant cross-contamination of products or reagents was observed between the individual runs. This approach could in general open a widely applicable methodology for the flexible and diversified synthesis of solvent and metal-free fine chemicals and pharmaceuticals on various scales, which is schematically outlined in Fig. 1.

3
Catalysis in PEG/CO$_2$ Biphasic Systems

The introduction of the PEG side chains in ligand **1** increased the efficacy of the recycling process described in Sect. 2 as compared to the parent ligand triphenylphosphine, by further enhancing the retention of the catalyst during extraction. The choice of PEG was motivated by its established limited solubility resulting from high polarity and low volatility at sufficiently large molecular weight [13]. Therefore, PEG and scCO$_2$ form a biphasic liquid/supercritical system, where catalysts can be dissolved in the polar liquid phase and the substrates/products partition into the supercritical phase.

The utilisation of PEG/CO_2 systems in organometallic catalysis was reported for the first time for the hydrogenation of styrene to ethylbenzene in the presence of Wilkinson's catalyst $RhCl(PPh_3)_3$ [14]. Although they are waxy solids at room temperature, PEGs with molecular weights M_n of 900 (or higher) are liquids at 40 °C when exposed to CO_2 at elevated pressure. As a result, biphasic systems consisting of PEG-900 or PEG-1500 and scCO$_2$ were

Table 2 Biphasic catalytic oxidation of alcohols using PEG-stabilised palladium nanoparticles in scCO$_2$[a]

Substrate	Product	T [°C]	t [h]	Run	Conversion [%]	Selectivity [%]
(5)	(6)	65	1.5	1	81.0	99.8
				2	99.2	99.3
				3	100	99.1
		65	1.5	2	100	98.1
		65	1.5	2	99.8	99.2
		80	13	1	83.1	99.8
				2	96.2	98.8
				3	99.8	97.5
				4	100	98.7
		80	13	2	56.9	98.8
		80	26	2	99.5	98.9
		80	13	2	45.8	95.5
		80	4	1	65.5	57.5[b]

[a] General reaction conditions: palladium cluster **7** (0.1 mmol Pd), PEG-1000 (2.40 g), substrate (1.99 mmol), $d(CO_2/O_2) = 0.55$ g/mL
[b] Butyric acid butyl ester is formed as a second product together with small amounts of butanal

found to be well suited as reaction/extraction systems for the hydrogenation of ethylbenzene, with neither detectable loss of catalyst activity (> 99%) in each of five consecutive cycles nor detectable amounts of Rh metal present in the product (< 1 ppm).

We showed that the application of PEG/CO$_2$ biphasic catalysis is also possible in aerobic oxidations of alcohols [15]. With regard to environmental aspects it is important to develop sustainable catalytic technologies for oxidations with molecular oxygen in fine chemicals synthesis, as conventional reactions often generate large amounts of heavy metal and solvent waste. In the biphasic system, palladium nanoparticles can be used as catalysts for oxidation reactions because the PEG phase both stabilises the catalyst particles and enables product extraction with scCO$_2$.

As a test reaction the oxidation of 3-methyl-2-buten-1-ol (**5**) to 3-methyl-2-buten-1-al (**6**) was chosen using the palladium cluster [Pd$_{561}$(phen)$_{60}$(OAc)$_{180}$] (**7**) (phen = 1,10-phenanthroline). A dispersion of **7** in PEG-1000 together with **5** was charged into a high-pressure reactor, which was subsequently pressurised with CO$_2$ and O$_2$. Both the substrate and the resulting product were predominantly contained in the CO$_2$ phase, as elucidated by online GC monitoring. Oxidation occurred efficiently to the corresponding aldehyde yielding **6** with quantitative conversion (99% selectivity). The product isolated by scCO$_2$ extraction contained only 2.3 ppm of Pd according to atomic absorption spectroscopy. Recycling of the catalyst was possible by simply recharging the reactor with fresh substrate, and catalyst activity was fully retained. A variety of alcohols were tested for oxidation, the results being shown in Table 2.

The promising results obtained under batch-wise operation could be confirmed also under continuous-flow conditions. As schematically shown in Fig. 2, the substrate was injected with an HPLC pump into a stream of premixed CO$_2$ and O$_2$ before entering the reactor. The supercritical mixture was bubbled through the stationary PEG phase containing the Pd nanoparticles.

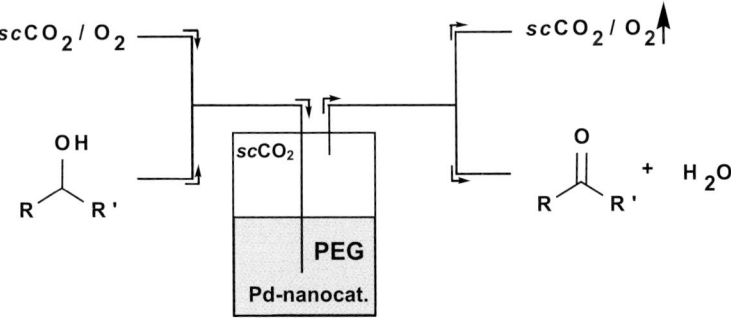

Fig. 2 Schematic representation of continuous-flow aerobic oxidation of alcohols using Pd nanoparticles in a PEG/scCO$_2$ biphasic system

The products were extracted with the mobile $scCO_2$ phase and isolated by controlled expansion after leaving the high-pressure reactor. A stable performance was achieved over more than 40 h time on stream with this simple setup.

4
Catalysis in Ionic Liquid/CO₂ Biphasic Systems

Ionic liquids (ILs) have attracted considerable interest in recent years as advantageous solvents being practically non-volatile, exhibiting a large structural variety, and showing the ability to dissolve a remarkable variety of organic compounds and organometallic complexes, which makes them interesting solvents for homogeneous catalysis [16–20]. In an IL/scCO₂ biphasic system, the application of ILs as solvents for catalytically active metal complexes can be beneficial, while $scCO_2$ can be used as solvent for substrates and products as well as extraction medium and for fine-tuning the reaction parameters. The concept of this two-phase approach relies on the fact that $scCO_2$ is highly soluble in ILs, while typical ILs are insoluble in $scCO_2$ owing to their negligible volatility [21]. If the organic compounds are readily soluble in $scCO_2$, it is possible to selectively extract dissolved organic matter from IL solutions with nearly quantitative recovery of the isolated material. Consequently the application of IL/scCO₂ systems is a promising tool for biphasic catalysis.

The concept has been used in recent years by a number of groups for reactions such as hydrogenation and hydroformylation [6, 7]. As an example, tiglic acid can be hydrogenated asymmetrically with quantitative conversion and high enantiomeric excesses (ee) using $Ru(O_2CMe)_2((R)\text{-tolBINAP})$ as catalyst in the IL 1-n-butyl-3-methylimidazolium hexafluorophosphate ([BMIM][PF$_6$]) as solvent [22]. Extraction with $scCO_2$ proved to be successful as the catalyst is much more soluble in the IL than in $scCO_2$, yielding the product in essentially pure form. The catalyst remained equally active in four consecutive runs.

Another example of hydrogenation was provided by transforming dec-1-ene and cyclohexene to the corresponding saturated products using Wilkinson's catalyst [RhCl(PPh$_3$)$_3$] in the presence of [BMIM][PF$_6$] as the solvent [23]. Both substrates could be hydrogenated with high conversions and were extracted from the reaction mixture with $scCO_2$. Recyclability experiments were carried out for dec-1-ene by recharging the reactor with substrate and carrying out the reaction again, with conversions reaching ca. 98% in each of four consecutive runs.

Hydroformylations were also carried out in IL/scCO₂ biphasic reaction systems [7]. In particular, it was demonstrated that continuous-flow systems could be operated successfully for the Rh-catalysed hydroformylation of hex-1-ene [24] and 1-dodecene [25].

Chiral secondary amines are interesting synthetic targets with strong relevance for industrial applications in fine chemistry. An attractive route to obtain these products is the asymmetric hydrogenation of imines. As a prototypical example of the development of biphasic IL/scCO$_2$ reaction mixtures, we investigated the performance of cationic iridium complexes as catalysts for the hydrogenation of N-(1-phenylethylidene)aniline (**8**) to give (R)-phenyl-(1-phenylethyl)amine (**9**) (Scheme 2) [13].

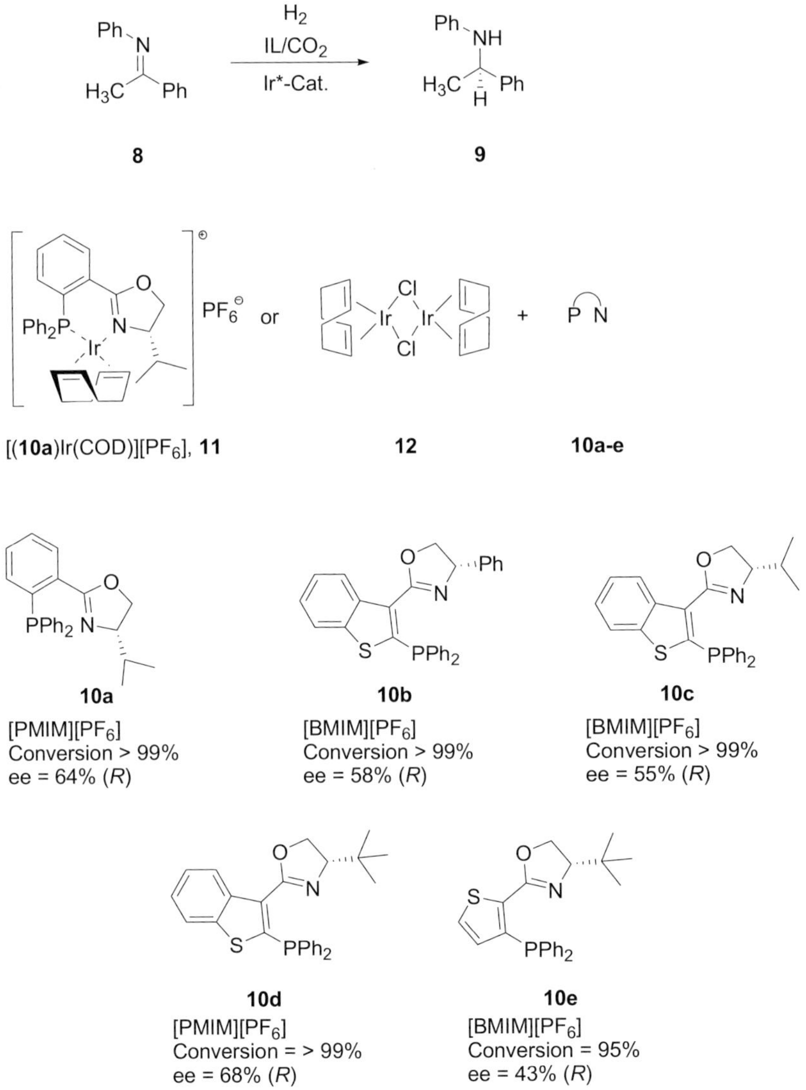

8   **9**

[(**10a**)Ir(COD)][PF$_6$], **11**   **12**   **10a-e**

10a
[PMIM][PF$_6$]
Conversion > 99%
ee = 64% (R)

10b
[BMIM][PF$_6$]
Conversion > 99%
ee = 58% (R)

10c
[BMIM][PF$_6$]
Conversion > 99%
ee = 55% (R)

10d
[PMIM][PF$_6$]
Conversion = > 99%
ee = 68% (R)

10e
[BMIM][PF$_6$]
Conversion = 95%
ee = 43% (R)

Scheme 2 Enantioselective hydrogenation of N-(1-phenylethylidene)aniline (**8**) in IL/CO$_2$ and catalytic systems based on ligands **10a–e**

In a typical experiment, the appropriate IL (2.0 mL), the iridium complex **11** (3×10^{-3} mmol) and the substrate **8** (**11:8** = 500:1) were loaded under argon in a window-equipped stainless steel autoclave (V = 12 mL). The reactor was then pressurised with H_2 and the desired amount of CO_2, followed by heating under stirring to 40 °C for a standard reaction time of 22 h. The products were collected for GC and HPLC analysis by extraction of the IL phase with hexane after cooling and venting, or alternatively isolated by CO_2 extraction. Representative results are summarised in Table 3.

Comparing the results of entries 1–3 in Table 3, one can see that the reaction requires a pressure of about 100 bar of hydrogen to proceed to completion in the absence of CO_2. In contrast, a partial pressure of 30 bar of H_2 is sufficient in the biphasic IL/scCO₂ system. This striking difference results from the significant influence of the compressed CO_2 on the physico-chemical properties of the IL phase.

In particular, high-pressure NMR studies revealed that the additional CO_2 pressure leads to a marked increase of H_2 concentration in the ligand phase. This reflects the strongly non-ideal behaviour of this system, as the concentration of gases in a liquid would be proportional only to the partial pressures in an ideal mixture. Typical enhancement factors of 3–5 are observed for the hydrogen concentration from the increase of the signal of dissolved hydrogen in the liquid phase upon addition of CO_2 (Fig. 3). Similar observations have been made for other reactive gases [14] and other biphasic systems [15] with

Table 3 Enantioselective hydrogenation of N-(1-phenylethylidene)aniline (**8**) to (R)-phenyl-(1-phenylethyl)amine (**9**) in IL/CO₂ systems[a]

Entry	IL	Cat.	H_2 [bar]	CO_2 [g]	Conv. [%]	ee [%]
1	[EMIM][BTA]	**11**	30	–	3	–
2	[EMIM][BTA]	**11**[b]	100	–	97	58
3	[EMIM][BTA]	**11**	30	8.9	> 99	56
4	[4MBP][BTA]	**11**	30	–	> 99	53
5	[4MBP][BTA]	**11**	30	9.8	> 99	52
6	[PMIM][PF₆]	**11**	35	7.5	> 99	61
7	[PMIM][PF₆]	**12/10a**[c]	30	9.2	> 99	64
8	[BMIM][BF₄]	**11**	30	7.6	92	30
9	[EMIM][BARF][d]	**11**	30	8.9	> 99	78

[a] Standard conditions unless noted otherwise: **8** : Ir = 500 : 1, V(reactor) = 12 mL, V(IL) = 2.0 mL, T = 40 °C, t = 22 h
[b] **8** : Ir = 100 : 1
[c] **8** : **10a** : Ir = 250 : 1 : 1, in situ protocol
[d] [EMIM][BARF] = 0.7–0.8 g

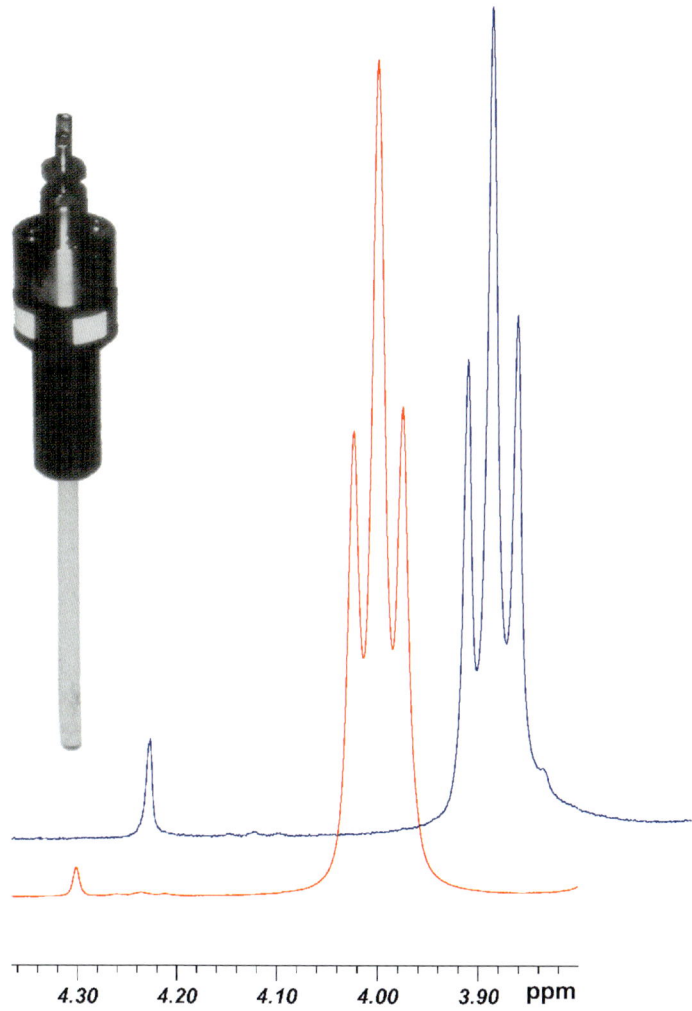

Fig. 3 High-pressure NMR tube and ^{1}H NMR spectrum of the IL [BMIM][BTA] under a pressure of 30 bar of hydrogen without CO$_2$ (lower trace) and in the presence of an additional 80 bar of CO$_2$ (*upper trace*). The signal at 4.3 ppm results from dissolved hydrogen, indicating the increase in solubility in the presence of CO$_2$ by the increase in intensity relative to the signal at 4.0 ppm from the IL solvent

different techniques. Additional effects, such as reduced viscosity of the IL, may also play a role [16].

The investigations showed that enantioselective hydrogenation of imines in IL/CO$_2$ mixtures is possible and the combination of the two phases is beneficial in the reaction step. Furthermore, quantitative extraction of the product with CO$_2$ from the reaction mixture could be established successfully. A simple separation of product and catalyst with reuse of catalyst is not possible for this reaction in conventional reaction media because of the instability of the active species. In the IL/scCO$_2$ system, however, recycling experiments showed stable conversion and no changes in enantiomeric excess over nine repetitive batches, demonstrating that the reaction protocol and the product isolation result in efficient catalyst immobilisation.

One of the most attractive features of the IL/CO$_2$ approach to homogeneous catalysis is the development of continuous processes [7]. Consequently it needs to be demonstrated that the combination of a suitable IL and compressed CO$_2$ can offer more potential for process optimisation than just a simple protocol for batch-wise catalyst recycling. As an example we were able to activate, tune and immobilise Ni catalyst **13** in a continuous-flow system for the hydrovinylation of styrene (Scheme 3). Styrene is co-dimerised with ethene yielding 3-substituted 1-butenes [26, 27]. We could show that this powerful carbon–carbon bond-forming reaction can be achieved with high enantioselectivity in batch-wise operation and in continuous-flow systems.

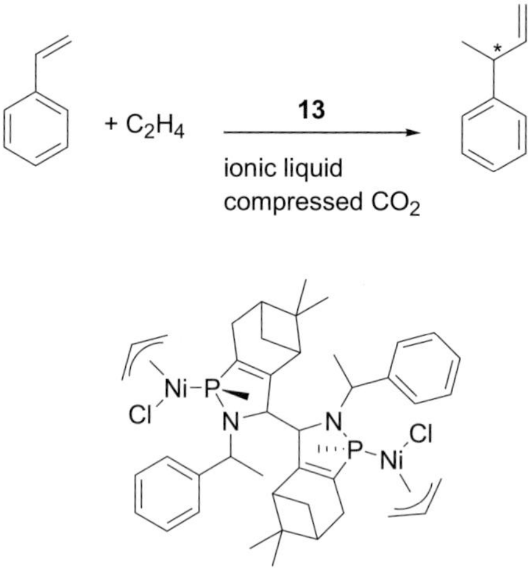

Scheme 3 Hydrovinylation of styrene with Wilke's catalyst **13** in an IL/scCO$_2$ biphasic system

First, suitable combinations of catalyst **13** and ILs were determined in batch experiments. Using [EMIM][BARF] as IL complete conversion was obtained with 89% ee of (R)-3-phenyl-1-butene. [EMIM][Tf$_2$N] gave lower ee values in the range of 65%. Nevertheless, better availability and easier handling made this the IL of choice for recycling and continuous-flow experiments. In batch-wise recycling experiments of catalyst **13** in ILs, it was found that the products could be readily isolated by extraction with scCO$_2$. However, the batch mode led to rapid deactivation of the catalyst within three to four cy-

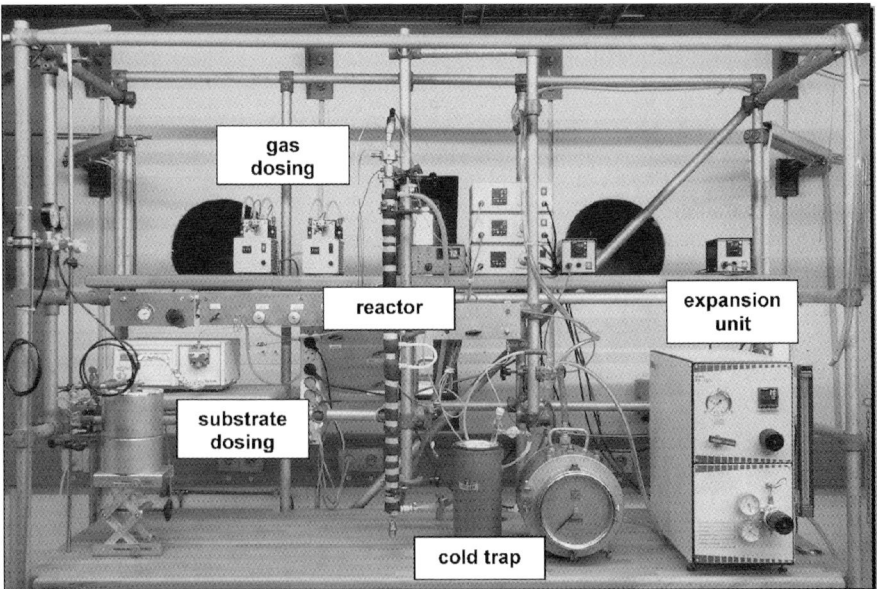

Fig. 4 Setup for continuous-flow asymmetric hydrovinylation using an IL/scCO$_2$ biphasic system. Liquid and gaseous substrates are mixed with the scCO$_2$ stream before entering the tubular reactor unit and bubbled through the catalyst-containing IL using a capillary. The CO$_2$ flow leaves the reactor on top and the product is collected in a cold trap after controlled expansion to ambient pressure

cles. Control experiments showed that this deactivation was mainly due to the instability of the active species in the absence of substrate. In a continuous process, this situation is avoided and one may still reach stable operating conditions. Indeed, this could be confirmed in the present case.

The continuous-flow apparatus used for homogeneous catalysis in IL/CO$_2$ systems in this study is depicted in Fig. 4. The ionic catalyst solution is placed into the reactor where it is in intimate contact with the mobile reaction phase. In a typical experiment, the reactor was filled under an argon atmosphere with 0.16 g (0.19 mmol) of **1** dissolved in 39 mL of [EMIM][Tf$_2$N] and cooled to 0 °C. The reactor pressure was maintained constant at 80 bar with a continuous flow of compressed CO$_2$ (the exit flow was adjusted to about 30 L/min). Note that the CO$_2$ is liquid rather than supercritical under these conditions. This does not affect the operating principle, however, owing to the similar behaviour in the near-critical region. Over more than 60 h, catalyst **13** showed a remarkably stable activity (75–80% single pass conversion) and the enantioselectivity dropped only slightly (approx. 65 to 60% ee) over the long reaction period. The results clearly indicate that even a highly sensitive catalyst can show excellent catalytic performance upon immobilisation under continuous product extraction with compressed CO$_2$.

5
Catalysis in Inverted CO_2/H_2O Biphasic Systems

The examples described in Sects. 2–4 all use carbon dioxide as the phase to dissolve the substrates and/or products. Under continuous operation, the CO_2 phase thus serves as the mobile phase. However, one may also envisage a so-called inverted scenario, where $scCO_2$ becomes the stationary catalyst phase and a second liquid phase contains substrates and products. This allows the processing of components that are not or only very poorly soluble in $scCO_2$. Furthermore, as it uses a liquid as continuous phase, energy-demanding compression cycles of the CO_2 phase are avoided. A necessary prerequisite is the use of a sufficiently "CO_2-philic" catalyst as outlined below.

The most attractive liquid phase for such an inverted system is water. The combination of the environmentally friendly solvent water and CO_2 makes this a particularly nice example of a "green" solvent system. The fundamental physico-chemical properties of this biphasic system have been studied in great detail [28]. Organometallic catalytic reactions have been described in $H_2O/scCO_2$ systems using the standard approach with a water-soluble catalyst in the liquid phase and CO_2 as solvent for the organic components. Under such conditions, the selective hydrogenation of olefinic $C = C$ double bonds or carbonyl $C = O$ double bonds was established using water-soluble ruthenium catalysts ($C = O$ hydrogenation) or rhodium and palladium catalysts ($C = C$ hydrogenation) [29]. To overcome mass transfer limitations which occur at the phase boundary, the addition of surfactants helps significantly in enhancing reaction rates, as was shown for the hydrogenation of styrene [30].

Inverting the catalyst and product phase, which will be denoted in the following reaction as the $scCO_2/H_2O$ biphasic system, opens the possibility to utilise very polar and hence hydrophilic substrates, as often encountered in fine chemicals and pharmaceutical production. To generate a sufficiently CO_2-soluble catalyst, one possibility is to introduce long perfluoroalkyl side chains into the ligand. These "CO_2-philic" groups are known to increase the solubility of organometallic catalysts by several orders of magnitude [31], ensuring a practically exclusive partitioning into the CO_2 phase of a $scCO_2/H_2O$ system. This concept was first validated in the Rh-catalysed hydroformylation of polar substrate **14** using the CO_2-philic ligand 4-H^2F^6TPP for the in situ generation of the active catalyst (Scheme 4) [32].

An aqueous solution of the reactant was pumped into an autoclave charged with the Rh precursor complex, the fluorinated ligand, the CO/H_2 mixture and CO_2. During the reaction the catalyst/$scCO_2$ phase and the substrate/water phase were intimately contacted in an emulsion-like mixture upon rapid stirring of the reactor contents. The substrate was converted completely to aldehydes **15a** and **15b** within 20 h, and depending on catalyst loadings the turnover numbers varied between 500 and 1720. The aqueous phase was isolated readily from the bottom of the reactor after phase separa-

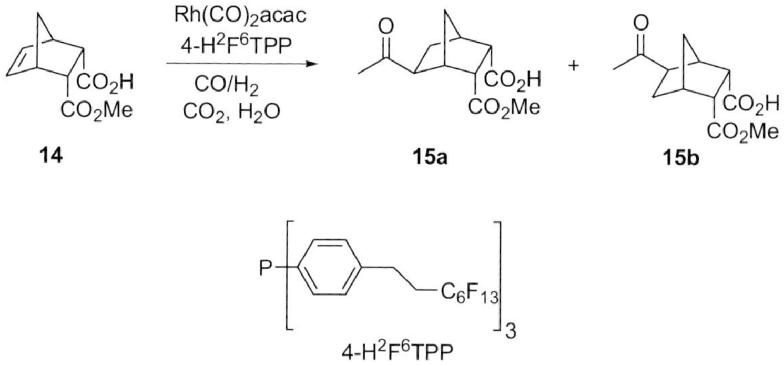

Scheme 4 Hydroformylation of 14 with scCO$_2$-soluble Rh catalyst in an inverted biphasic system scCO$_2$/H$_2$O

tion. The recycling of the catalyst met with only limited success, however, and rapid deactivation was observed. This may at least be partly associated with the low pH of the scCO$_2$/H$_2$O system, which is not the ideal environment for hydroformylation catalysts.

The Rh-catalysed hydrogenation of itaconic acid (16) (Scheme 5) to yield the corresponding saturated acid 17 was found to exhibit similar high activities, but much better long-term stability [33]. This reaction could be conducted in repetitive batch mode using the experimental setup schematically shown in Fig. 5. The catalyst used was generated in situ by employing [Rh(cod)$_2$]BF$_4$ (cod = cyclooctadiene) as catalyst precursor and 3-H^{2}F^{6}TPP as the ligand.

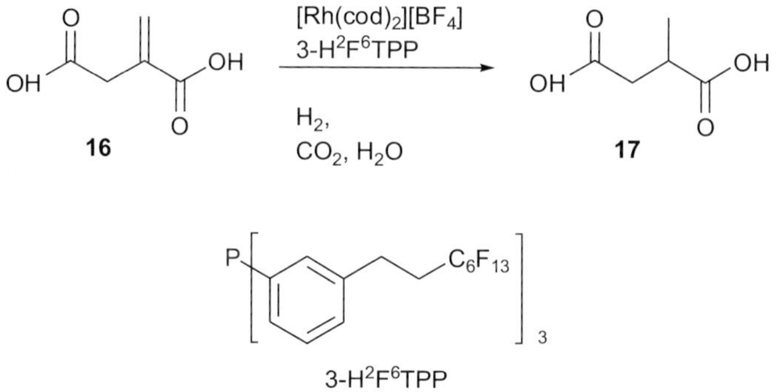

Scheme 5 Rhodium-catalysed hydrogenation of itaconic acid (16) in an inverted biphasic system scCO$_2$/H$_2$O

The catalyst was dissolved in scCO$_2$, hydrogen was added and this phase remained inside the reactor over the entire sequence of experiments. An aqueous solution of itaconic acid (16) was brought via an HPLC pump

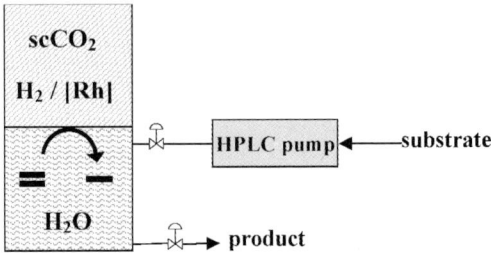

Fig. 5 Schematic representation of the experimental setup for hydrogenation in an inverted biphasic system $scCO_2/H_2O$

against the pressure into the autoclave. Stirring at 1000 rpm caused the two phases to mix intimately, but they separated immediately when stirring was stopped. To observe the reaction process, samples of the water layer were taken through a needle valve at the bottom of the reactor. After complete conversion and total removal of the product phase, a new batch of substrate was pumped into the reactor without depressurising the autoclave. Small amounts of CO_2 were added from time to time to compensate for the pressure drop resulting from the solubility of CO_2 in the removed aqueous phase.

The rhodium catalyst was recycled batch-wise four times. It was found that a short induction period occurred during the first reaction cycle. The following cycles showed a constant rate and no loss of activity was detected. A ligand-to-rhodium ratio of 5 : 1 led to a constant yield of 95% per cycle after 1 h. Within the four cycles a total turnover number of 1000 with a maximum turnover frequency of 234 h^{-1} was achieved. The leaching of rhodium and phosphorus into the aqueous layer was determined by inductively coupled plasma atomic emission spectrometry.[1] Rhodium leaching amounted to 14.2 ppm in the first run, then dropped to 3.6 ppm (second run) and reached values of 0.95 and 0.63 ppm in the third and fourth runs, respectively.

The CO_2-philic perfluoroalkyl-substituted (R,S)-3-H^{2}F^6-BINAPHOS ligand [34] was successfully applied to enantioselective hydrogenation in the inverted $scCO_2/H_2O$ system. The complex [Rh(cod)$_2$]BARF was chosen as metal source and the active catalyst was formed in situ. Using the same procedure as above, similar activities and more than 98% ee were obtained consistently over five subsequent cycles in the hydrogenation of methyl 2-acetamidoacrylate. The results demonstrate the potential of the inverted $scCO_2/H_2O$ system for asymmetric synthesis of chiral biologically active products.

---

[1] ICP measurements were determined by the group of Prof. Dr. A. Behr, Universität Dortmund.

6
Summary and Outlook

The approaches to catalyst immobilisation described in this chapter rely on the use of compressed, and in particular supercritical, carbon dioxide to separate products and catalysts on the basis of their different solubility behaviour in this medium. As typical organometallic catalysts often exhibit very limited solubilities in scCO$_2$, strategies to use scCO$_2$ as solvent for substrates and/or products are most obvious. In the present brief review, the use of scCO$_2$ as a "solubility switch" to precipitate homogeneous catalysts from neat reactions was highlighted and shown to lead to cartridge systems for flexible combination of various transformations. To establish continuous-flow processes, it was shown to be beneficial to combine scCO$_2$ with a second liquid phase. PEG and ILs were shown to be promising candidates in this context. Finally, an "inverted system" with scCO$_2$ as the catalyst phase and H$_2$O as the product phase was presented and discussed as a possible approach to asymmetric catalysis with hydrophilic substrates.

The examples illustrate that partitioning and separation into the stationary or mobile phase can be regulated and controlled by a combination of molecular design and reaction engineering concepts if scCO$_2$ is used in multiphase organometallic catalysis. This fascinating field of catalysis research lies at the interface of molecular sciences and process engineering, and its future development will require truly interdisciplinary efforts in both fields. Significant additional efforts in fundamental sciences are required to establish a methodological platform for industrial implementation in sustainable chemical manufacturing.

Acknowledgements Financial support of the Bundesministerium für Bildung und Forschung (BMBF), Max-Planck-Institut für Kohlenforschung and the Fonds der Chemischen Industrie is gratefully acknowledged.

References

1. Cornils B, Herrmann WA, Vogt D, Horvath I, Olivier-Bourbigon H, Leitner W, Mecking S (eds) (2005) Multiphase homogeneous catalysis. Wiley-VCH, Weinheim
2. Cole-Hamilton DJ, Tooze RP (eds) (2006) Catalyst separation, recovery and recycling. Springer, Dordrecht
3. Jessop PG, Leitner W (eds) (1999) Chemical synthesis using supercritical fluids. Wiley-VCH, Weinheim
4. Jessop PG, Ikariya T, Noyori R (1999) Chem Rev 99:475
5. Oakes RS, Clifford AA, Rayner CT (2001) J Chem Soc Perkin Trans 1, p 917
6. Leitner W (2002) Acc Chem Res 35:746
7. Cole-Hamilton DJ (2003) Science 299:1702
8. Leitner W (2003) Nature 423:930

9. Sellin MF, Cole-Hamilton DJ (2000) J Chem Soc Dalton Trans, p 1681
10. Kainz S, Brinkmann A, Leitner W, Pfaltz A (1999) J Am Chem Soc 121:6421
11. Fürstner A, Ackermann L, Beck K, Hori H, Koch D, Langemann K, Liebl M, Six C, Leitner W (2001) J Am Chem Soc 123:9000
12. Solinas M, Jiang J, Stelzer O, Leitner W (2005) Angew Chem Int Ed 44:2291
13. Kokova E, Petermann M, Weidner E (2004) Chem Ing Tech 76:280–284
14. Heldebrant DJ, Jessop PG (2003) J Am Chem Soc 125:5600
15. Hou Z, Theyssen N, Brinkmann A, Leitner W (2005) Angew Chem 117:2
16. Wasserscheid P, Welton T (eds) (2003) Ionic liquids in synthesis. Wiley-VCH, Weinheim
17. Holbrey JDS, Seddon KR (1999) Clean Prod Proc 1:2230
18. Welton T (1999) Chem Rev 99:2071
19. Wasserscheid P, Keim W (2000) Angew Chem Int Ed 39:3773
20. Gordon CM (2001) Appl Catal A 222:101
21. Blanchard LA, Hancu D, Beckman EJ, Brennecke JF (1999) Nature 399:28
22. Brown RA, Pollett P, McKoon E, Eckert CA, Liotta CL, Jessop PG (2001) J Am Chem Soc 123:1254
23. Liu F, Abrams MB, Baker RT, Tumas W (2001) Chem Commun, p 433
24. Sellin MF, Webb PB, Cole-Hamilton DJ (2001) Chem Commun, p 781
25. Webb PB, Sellin MF, Kunene TE, Williamson S, Slawin AMZ, Cole-Hamilton DJ (2003) J Am Chem Soc 125:15577
26. Jolly PW, Wilke G (1996) In: Cornils B, Herrman WA (eds) Applied homogeneous catalysis with organic compounds 2. Wiley-VCH, Weinheim, pp 1024–1048
27. RajanBabu TV, Nomura N, Jin J, Radetich B, Park H, Nandi M (1999) Chem Eur J 5:1963
28. Wiebe R (1941) Chem Rev 29:475
29. Bhanage BM, Ikushima Y, Shirai M, Arai M (1999) Chem Commun, p 1277
30. Jacboson GB, Ted Lee C Jr, Johnson KP, Tumas W (1999) J Am Chem Soc 121:11902
31. Kainz S, Koch D, Baumann W, Leitner W (1997) Angew Chem Int Ed 36:1628
32. McCarthy M, Stemmer H, Leitner W (2002) Green Chem 4:501
33. Burgemeister K, Franciò G, Hugl H, Leitner W (2005) Chem Commun, p 6026
34. Franciò G, Wittmann K, Leitner W (2001) J Organomet Chem 621:130

Top Organomet Chem (2008) 23: 109–147
DOI 10.1007/3418_041
© Springer-Verlag Berlin Heidelberg
Published online: 29 July 2006

Evaluation of Supercritical Carbon Dioxide as a Tuneable Reaction Medium for Homogeneous Catalysis

Stephan Pitter (✉) · Eckhard Dinjus · Cezar Ionescu · Constantin Maniut ·
Piotr Makarczyk · Florian Patcas

Institute for Technical Chemistry, Division of Chemical-Physical Processing (ITC-CPV),
Forschungszentrum Karlsruhe GmbH, PO Box 3640, 76021 Karlsruhe, Germany
stephan.pitter@itc-cpv.fzk.de

Abstract The present contribution highlights the relationship between scCO$_2$ properties, its solubilization power, and its use as a reaction medium for homogeneous catalysis. Current research activities under the lighthouse project "Smart Solvents, Smart Ligands" are presented, the focus being on criteria of conducting catalyzed processes in future applications.

Keywords Supercritical carbon dioxide · Solubility measurement ·
Homogeneous catalysis · Multi-phase catalysis · Hydroformylation

Abbreviations

ρ	Density
CESS	Catalysis and extraction using supercritical solutions
cod	1,5-cyclooctadiene
ConNeCat	Competence network catalysis
EOS	Equation of state
hfacac	Hexafluoracetylacetonate
ISI	ISI Web of Knowledge
p_c	Critical pressure
RTIL	Room-temperature ionic liquid
S	Selectivity (in Figs.)
S	Substrate(s) (in tables)
sc	Supercritical
$scCO_2$	Supercritical carbon dioxide
SCF	Supercritical fluid
Tc	Critical temperature
TOF	Turn-over frequency
XANTPHOS	4,5-Bis-diphenylphosphanyl-9,9-dimethyl-9,9a-dihydro-4aH-xanthene

1
Introduction

In the 21st century, the chemical industry will be increasingly influenced by environmental concerns with respect to waste treatment. Consequently, one of the aims of green chemistry is the replacement of hazardous materials (solvents, reagents) by less hazardous substances [1–3]. In order to comply with the requirements of a sustainable development, modern "state-of-the-art" chemistry should comprise:

• Efficient chemical processing
• Environmental protection
• Economic viability

Therefore, particular attention is paid to the use of volatile organic solvents in the manufacturing and processing of chemical products [4]. The use of water might solve some of these problems, but is certainly not suited for many important applications, because it might also result in large amounts of hazardous aqueous waste requiring further treatment. Due to its environmental advantages, supercritical carbon dioxide ($scCO_2$) has been investigated widely with regard to the replacement of organic solvents in chemical reactions [5]. $scCO_2$ is nontoxic, does not form air-polluting daughter products, and is relatively cheap and plentiful. CO_2 for supercritical fluid technologies is available in large quantities as a by-product of fermentation, combustion, and ammonia synthesis. Its price depends on the quantity delivered, as transportation is the main cost factor because carbon dioxide is available at a very low cost in various chemical plants (mainly ammonia plants).

The physical-chemical properties of a supercritical fluid are between those of liquids and gases: supercritical fluids (SCFs) indicate the fluid state of a compound in pure substance or as the main component above its critical pressure (p_c) and its critical temperature (T_c), but below the pressure for phase transition to the solid state, and in terms of SCF processing, a density close to or higher than its critical density.

This definition cannot be applied directly to mixtures, as phase equilibria of mixtures can be very complex. Nevertheless, the term supercritical is widely accepted because of its practicable use in certain applications [6]. Some properties of SCFs can be simply tuned by changing the pressure and temperature. In particular, density and viscosity change drastically under conditions close to the critical point. It is well known that the density-dependent properties of an SCF (e.g., solubility, diffusivity, viscosity, and heat capacity) can be manipulated by relatively small changes in temperature and pressure (Sect. 2.1).

Currently, application of $scCO_2$ in industry is an accepted and established technology. There are several applications of $scCO_2$, but among the most important are industrial extraction [7–10] and material processing [11–15]. In addition to being environmentally benign, the low viscosity of $scCO_2$ allows for a rapid mass transport not only in extraction processes but may be also of benefit in chemical reactions [16, 17]. If a chemical reaction is particularly fast, then mass transfer could be rate-limiting, and an increase in the diffusivity could lead to an enhanced reaction rate. In the absence of a phase boundary in the supercritical state, in addition, no interfacial mass transfer takes place. The gas-like mass transport characteristics of an SCF and its complete miscibility with permanent gases, coupled with the liquid-like ability to dissolve organic compounds of low to medium polarity, make it an ideal medium for mass transfer-limited reactions with gaseous reactants.

In catalysis applications, the tunable solvent properties result in a variety of effects, such as controllable component and catalyst solubilities. Moreover, it is possible that kinetic rates are affected by both temperature and pressure effects, equilibrium constants are shifted in favor of the desired products, and selectivity and yields are increased by manipulating the solvent's dielectric constant or by controlling the temperature in highly exothermic reactions through an adjustment of the solvent's heat capacity [18–23].

In view of the environmental concerns, $scCO_2$ represents a more environmentally friendly alternative to the traditional solvents. The general lack of reactivity could be essential to the future success of $scCO_2$ in replacing more conventional solvents. Even though $scCO_2$ has been touted as a modern remedy of many commercial problems, the use of CO_2 as a solvent is complicated

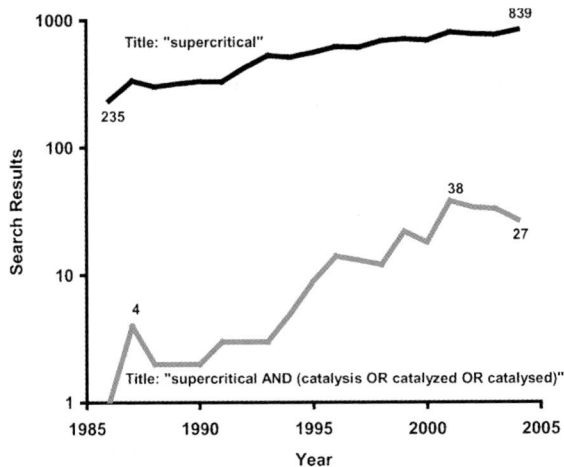

Fig. 1 Scientific publications referenced by ISI

due to the low solubility of many reactants even under supercritical conditions. Many industrial applications are hindered by this as well as by the fact that high-pressure equipment requires additional investment. This is the reason why only very few developments have resulted in new applications, although the number of reports dealing with SCFs (Fig. 1) is constantly rising.

In particular, catalytic and analogous reactions in SCFs will become increasingly attractive because of the environmental legislation. To implement future processes based on the use of $scCO_2$ as medium for catalysis, representative data need to be provided, which describe the states present under reaction and separation conditions. Such data are crucial to prevent losses of catalysts and/or expensive ligands and to achieve the desired product purities [24].

In this chapter, brief information shall be given on the relationship between $scCO_2$ properties, its solubilization power and its use as reaction medium for homogeneous catalysis. Current research activities from our laboratories [25] together with selected examples from the literature shall be presented, the focus being on making available criteria as to how to conduct those processes in future applications.

2
Solvent Characteristics of Carbon Dioxide

2.1
Fundamental Aspects

The critical point of CO_2 is readily accessible at 304.1282 K and 7.3773 MPa. The supercritical state of CO_2 is characterized by properties that are partly similar to liquids and partly similar to gases [6]. Figure 2 represents the pressure-density isotherms in the range from 280 to 720 K. When changing the pressure, it is obvious that no sharp increase/decrease of the density takes place above the critical point due to the presence of the single-phase state. Therefore, changes in temperature and/or pressure allow for a fine tuning of the CO_2 density. The largest variation of solvent density is attained in the vicinity of the solvent's critical point, where the solvent compressibility is large and small changes in pressure yield large changes in density [26]. In the range of technically applicable pressures between 5 and 30 MPa, the density is somewhat between the gaseous and the liquid state.

Figures 3 to 5 shows the density, heat capacity, and viscosity as a function of the pressure near the critical pressure of CO_2 at 320 K. At lower and higher pressures, the curves exhibit the typical behavior of gas and liquid, respectively. Above the critical pressure, the viscosity follows the ascent of density.

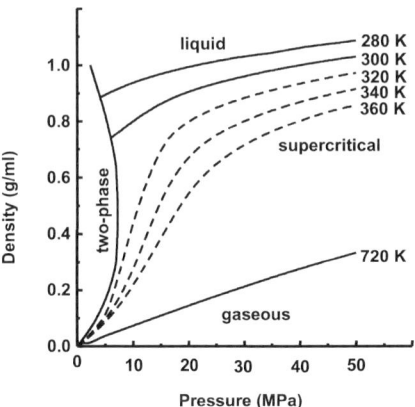

Fig. 2 Pressure–density isotherms of CO_2

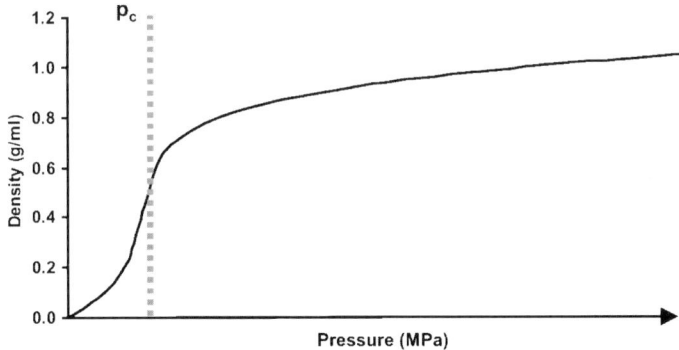

Fig. 3 Pressure-density isotherm of CO_2 at 320 K

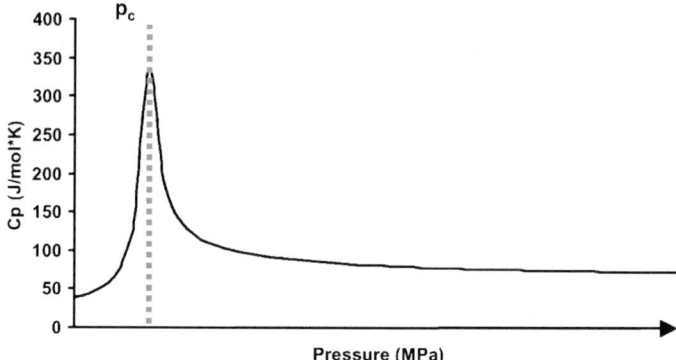

Fig. 4 Pressure-heat capacity isotherm of CO_2 at 320 K

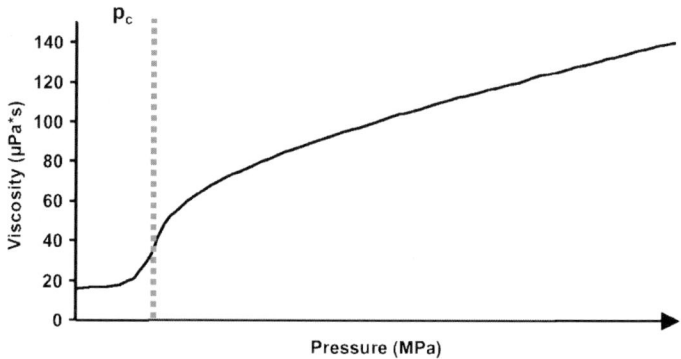

Fig. 5 Pressure-viscosity isotherm of CO_2 at 320 K

By theory, heat capacity is infinite at the critical point, which has also been proved experimentally.

Both viscosity and diffusivity change from gas-like to liquid-like with increasing pressure: Viscosity and diffusivity (which is strongly related to viscosity) are also temperature- and pressure-dependent and, in general, an order of magnitude lower and higher at least than in the liquid phase, respectively.

Carbon dioxide is a non-polar solvent characterized by a low polarizability per volume, a low Hildebrand solubility parameter, and a low dielectric constant. The dielectric constant of CO_2 as a function of pressure is shown in Fig. 6 [27].

These values are similar to those of saturated organic solvents, thus indicating similar solvent properties. The drastic change of density with pressure near the critical points is also obvious, but the change in the absolute value is small. Due to these properties, it was commonly accepted that only volatile

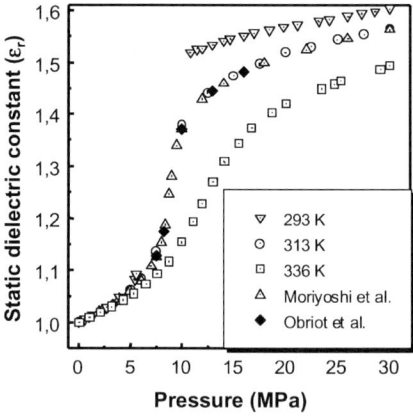

Fig. 6

or relatively non-polar compounds are soluble in CO_2. On the other hand, the CO_2 molecule possesses a significant quadrupole moment and the ability to act as a weak Lewis base (e.g., in interaction with carbonyl groups) or a weak Lewis acid (or to form intermolecular hydrogen bonds) [28]. The polarity of the C,O double bond causes interactions with polar molecules like water or RTILs (room-temperature ionic liquids) [29, 30]. Consequently, $scCO_2$ was described as "quadrupolar solvent" [31] and its solubility properties are not generally predictable.

The solvent power of $scCO_2$ mainly depends on its density, the temperature, and the intermolecular interaction between the solvent and the solute:

- Solubility increases with density (or pressure) at constant temperature;
- solubility may increase or decrease when temperature is raised at constant pressure; heating a supercritical solution from T_c under constant pressure initially leads to a decreasing solubility, as the density of the fluid decreases; further heating results in increased solubility, because the vapor pressure of the solute rises proportionately faster than the density of the fluid falls; in contrast to this, heating a solution under constant volume does not change the solubility, because the density remains constant.

As a result of the occurring density increase, the partial molar volume of solutes is often large and negative. The role of clustering, also around neutral solutes in chemical reactions, was reviewed by Tucker and Kajimoto [32, 33]. The solvent power of $scCO_2$ resulting from density can be altered by more than an order of magnitude by small changes in temperature or pressure, as outlined above. Reduction of pressure generally results in a lower solubility up to a complete separation of the solutes from the solvent.

2.2
Methods for the Determination of Solubilities in scCO$_2$

Various procedures have been reported in literature to investigate the solubility of compounds in scCO$_2$ [34]. They can be divided into static [34, 35] and dynamic [36, 37] methods. Most recently, also parallel techniques have been developed for solubility measurements [38]. All methods, except for the static synthetic one (see below), require analytical measurements to quantify the amount of solute dissolved in scCO$_2$. Mainly spectroscopic, chromatographic, and gravimetric techniques are applied for the analytical measurements.

According to the type of the vessel used, two subcategories of static methods are distinguished. In the case of analytic methods [39], a constant volume equilibrium cell is applied, where the compound under test is kept in contact with a well defined amount of scCO$_2$. After achieving the solvatization equilibrium, sampling is applied to determine the solubility. The second static method is commonly termed synthetic [40]. Solubility is often determined by using a variable volume cell, thus allowing for the adjustment of the operating conditions. The measurements are carried out at constant temperature by changing the pressure of a single phase mixture, until phase separation occurs. The vessel is usually equipped with a sapphire window to allow for the visual observation of the separation or cloud points.

With the dynamic methods, the scCO$_2$ flows continuously on or through the compound. The operating conditions are chosen to ensure that the outlet flow is in solvatization equilibrium. Solubility of scCO$_2$ in the component is usually neglected.

2.3
Prediction of the Solubility in scCO$_2$

The solubility of substances in SCFs has been described by many different approaches [41–43]. Based on experimental data, theoretical treatment allows for modeling the solubility in scCO$_2$ [44, 45]. Other approaches are based on equations of state (EOS) or on statistic models [46, 47].

To model the solubility of a solute in an SCF using an EOS, it is necessary to have critical properties and acentric factors of all components as well as molar volumes and sublimation pressures in the case of solid components. When some of these values are not available, as is often the case, estimation techniques must be employed. When neither critical properties nor acentric factors are available, it is desirable to have the normal boiling point of the compound, since some estimation techniques only require the boiling point together with the molecular structure. A customary approach to describing high-pressure phenomena like the solubility in SCFs is based on the Peng-Robinson EOS [48, 49], but there are also several other EOS's [50].

Statistical methods correlate the solubility with the density, pressure, and temperature. For example, Chrastil et al. adopted a semi-empirical model for the calculation of the solubility from $scCO_2$ density and temperature and, hence, of the number of solvent molecules participating in the solvatation [51].

2.4
Effects of the Molecular Structure of the Solute on its Solubility in scCO$_2$

Generally, the solubility characteristics of organic compounds depend on several properties of the participating components. For the solute, these properties are the molecular size and structure, polarity, dipole moment, vapor/sublimation pressure, and, in the case of a solid solute, also its melting characteristics. When using $scCO_2$ as the solvent, mainly its dipole moment and quadrupole moment influence the solvatation process (Sect. 2.2).

Low molecular weight hydrocarbons are moderately to well soluble. Naphthalene is widely used as standard compound for solubility measurements in $scCO_2$. Some general trends for the solubility of organic compounds are as follows:

- Saturated hydrocarbons are completely miscible with $scCO_2$ in the pressure and temperature range of technical interest, if the number of carbon atoms is smaller than 13;
- the solubility of small to middle-sized olefins rises with the number of carbon atoms;
- aromatics are typically less soluble than alkanes of similar molecular weight;
- the solubility of alcohols shows a trend similar to that of olefins. Up to n-hexanol, alcohols are completely miscible with $scCO_2$, but longer-chained alcohols are only partly soluble;
- nitrogen-containing molecules differ in their solubility. Pyridine and picoline are fully miscible with $scCO_2$, but aniline has a solubility of 3 mass % only. Interestingly, pyrolidine forms with CO_2 an insoluble solid aggregate at higher pressures.

Also several mixed solute systems have been reported, mostly with respect to SCF extraction [52–54]. However, CO_2 is a relatively poor solvent for polar organics and many other solutes. The chemical modification of the solute has been exploited to influence its solubility in $scCO_2$, which is mainly used in extraction applications and catalysis processes. A strategy that has been found to be quite successful in overcoming the limitation of poor solubility is to make use of CO_2-philic functional groups. In particular, fluorinated alkyl [55, 56], sugar [57, 58], and alkyl siloxane [59] groups have proven to be useful for dissolving compounds in $scCO_2$ [60]. The first method applied and, thus, the by far most widespread method is the introduction of fluorocarbon substituents. Most fluorocarbons exhibit high solubilities in liquid and scCO$_2$ and, generally, fluorinated compounds are much more soluble in CO$_2$ than their hydrocarbon counterparts. Contradicting theories of the nature of CF$_n$/CO$_2$ interaction, based on both experimental and theoretical approaches [45, 61–65], have been given, but the nature of this increased solubility is not fully understood [66]; an increase in solute–solvent van der Waals interactions, as the fluorination degree is increased, combined with a lower solute–solute attraction, may account for this. Whilst the interaction energies of the complexes of CO$_2$ with fluorocarbons and hydrocarbons are comparable, the fundamental nature of their interactions is different. While CO$_2$ acts as a weak Lewis acid in CO$_2$-fluorocarbon interactions, it acts as a weak Lewis base in CO$_2$-hydrocarbon interactions.

CO$_2$-philic molecules have been utilized for the design of metal-mobilizing ligands to be used in scCO$_2$ [67–69, 135–137], e.g., as shown in Fig. 7a [55] and for the synthesis of surfactants that form micelles, emulsions, and micro emulsions in CO$_2$, e.g., as shown in Fig. 7b. [70] Polymer solubility in scCO$_2$ has been studied [71] and utilized for polymer synthesis [72–74]. Recently, DeSimone and co-workers synthesized high-molar-mass fluoropolymers in scCO$_2$, and studied the polymerization kinetics [75].

As highly fluorinated compounds are expensive and the biodegradability is low, there is a great incentive to design new, low-cost CO$_2$-philic groups particularly for applications that involve the processing of large volumes of waste. The specific interaction of CO$_2$ molecules with Lewis base groups, especially carbonyl groups, has also been utilized in the design of CO$_2$-philic materials: Beckman et al. synthesized hydrocarbon-based, carbonyl-supported, poly(ether-carbonate) copolymers (Fig. 8) soluble in liquid CO$_2$ by maximizing the entropic and enthalpic contributions to solvatation [76].

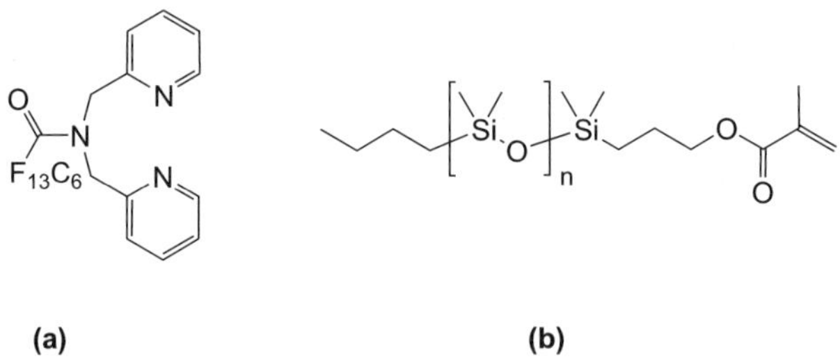

(a)   (b)

Fig. 7

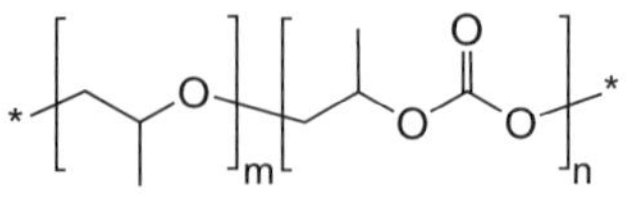

Fig. 8

3
Solubility of Metal Complexes in CO_2

3.1
General Aspects of Metal Complex Solubility in $scCO_2$

The tuneable solvent capability of $scCO_2$ offers the potential for a subtle control of reactions in order to achieve higher selectivities and improved reaction rates. Furthermore, the separation of extractives or, in the case of a synthesis, of reactants, products, and catalysts by simple decompression could be facilitated. The low solubility of many metal complexes and catalysts usually is an obstacle to their exploitation in $scCO_2$-based processes. For instance, the solubility of a homogeneous catalyst needs to be sufficiently high to ensure participation of all active metal centers during a catalyzed reaction. In particular for reactions, solubility properties are difficult to predict, because the component composition is continuously changed with conversion.

The solubility of many metal complexes is considerably limited due to their high polarity. As reported in literature, the solubility-promoting factors in organometallic and coordination metal complexes can be summarized as follows:

Complexes preferably need to be non-polar, uncharged, and hydrophobic.

The use of modifiers, either by addition or in terms of a chemical modification, is applicable for overcoming the limitation of low complex solubility.

Added modifiers, such as methanol, can affect the solubility of a compound in two different ways. They can make the $scCO_2$ a more polar medium that is better able to dissolve the metal complex or they coordinate to the metal center, resulting in a decrease of the overall polarity of the complex. A combination of both effects is also possible. It should be emphasized that by adding reagents or cosolvents to $scCO_2$, its properties can change significantly. Although CO_2 is described as supercritical in many literature examples, this may not be the case for the overall reaction mixture with significant concentrations of additional reagents. Often the term "dense-phase CO_2" is used, particularly when there is some uncertainty regarding the actual phase behaviour of a mixture.

Direct enhancement of metal complexes through chemical modification is based on the introduction of CO_2-philic functional groups. As reported by Wai et al., for example, metal-chelating ligands are made soluble by attaching CO_2-philic tails to the ligands in order to extract metal contaminants from soils and sludges with $scCO_2$ as the extraction solvent [77]. This technique is also applicable for catalytically active metal complexes [135–137]. The use of CO_2-philic moieties in ligand syntheses has significantly increased in the last decade. Perfluorinated alkyl [78–80], carbonylated [81] or silyl groups [82, 83], directly attached to the ligand backbone (Fig. 9), have allowed for the design of homogeneous catalysts, analogous to the conventional ones.

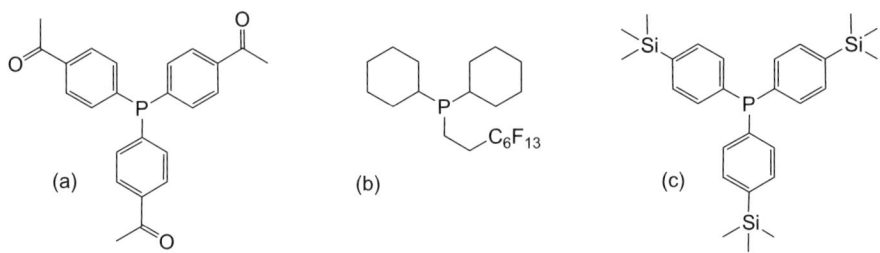

Fig. 9 Examples of CO_2-philic ligands for solubilizing metal complexes and catalysts; **a** refer to [81]; **b** refer to Inorg Chem 38:5277; **c** refer to [83]

The high solubility of ferrocene derivatives was attributed to the nature of bonding, in which the metal orbitals are filled and no free coordination sites are available for interaction with the solvating CO_2, thus resulting in a system having properties similar to those of an aromatic compound [84, 85]. Also some metal carbonyl complexes were shown to have moderate solubility in $scCO_2$, as reported by Walther et al. [86]. Soluble complexes have been applied to many important catalytic reactions in $scCO_2$: Olefin hydrogenation [73, 87–93, 137]; hydroformylation (Sect. 5); Heck-, Stille-, and Suzuki-coupling [59, 94–98]; olefin metathesis reactions [99, 100]; alkoxycarbonylation [101]; Pauson-Khand reaction [102]; hydrovinylation [103]; cyclotrimerization [104, 105]; olefin polymerization [106, 107]; hydroboration [108]; hydroarylation [109]; hydroaminomethylation [110]; olefin oxidation [111, 112]. There are also examples of utilizing $scCO_2$ as solvent and substrate; e.g., CO_2 hydrogenation with non-fluorous soluble catalysts [113, 114] and vinylcarbamate and N-formylmorpholine synthesis [115, 116];

A significant drawback in particular of the widely employed fluoroalkyl groups is their somewhat costly synthesis which often requires multiple steps and changed donor properties that might affect the activity of a catalyst. A successful solution of the latter problem is using an appropriate spacer. For the reasons as outlined above, the design of new solubilizers, that are straightforward to prepare, but do not modify the chemical properties in complex formation is a highly desirable target.

One key limitation to introducing metal complexes in $scCO_2$-based processes is the understanding of the CO_2-philicity on a fundamental theoretical basis. Furthermore, systematic solubility studies in $scCO_2$ are required for a rational catalyst design. In order to design and optimize reaction and separation processes, also solubility and phase equilibrium data with respect to "real" phase compositions are necessary. Up to now, there has been a significant lack of investigations dealing with metal complexes, also because such data are difficult to obtain experimentally, as many metal complexes are sensitive, decompose at higher temperatures or react under "real" conditions.

3.2
Solubility of $Co_2(CO)_6$(Phosphine)$_2$ in Pure scCO$_2$

Systematic solubility measurements of organometallic transition metal pre-catalysts were performed in our laboratories as part of the ConNeCat activities [25]. These activities focus on a multi-phase approach, in which product separation is to be achieved by a thermally or pressure-controlled phase separation, while the catalyst is completely precipitated.

Modified cobalt complexes of the type $trans$-$Co_2(CO)_6$(phosphine)$_2$ are promising candidates for certain transition metal-catalyzed reactions, in particular for the hydroformylation of long-chained olefins [117]. A series of complexes $Co_2(CO)_6[P(alkyl)_n(aryl)_m]_2$ (n: 0, 1, 2, 3; m: 3 − n) was synthesized and used for solubility measurements. Since the basicity of phosphines affects the catalytic activity, use of fluorous substituents might induce unexpected changes in the activity. Therefore, also derivatives with an additional ethyl spacer between the fluorous group and the phosphine moiety were examined (Sect. 3.1).

The Grignard reaction is one general method for the access to fluorinated phosphines [118]. Fluorinated phosphines are also related to catalysis applications in fluorous reaction media [119–121]. As a first example, $trans$-$Co_2(CO)_6[P(i$-Pr$)_2\{3,5$-$bis(CF_3)$-$C_6H_3\}]$ (**5b**) was synthesized in two steps by the reaction of $3,5$-$bis(CF_3)$-C_6H_3MgBr with $(i$-Pr$)_2$PCl to **5a** with a yield of 90%, followed by the complexation of $Co_2(CO)_8$ in refluxing benzene (Fig. 10).

In a first approach [117], an extraction apparatus was constructed as shown in Fig. 11 to determine the solubility of **5b** in scCO$_2$ by a dynamic method.

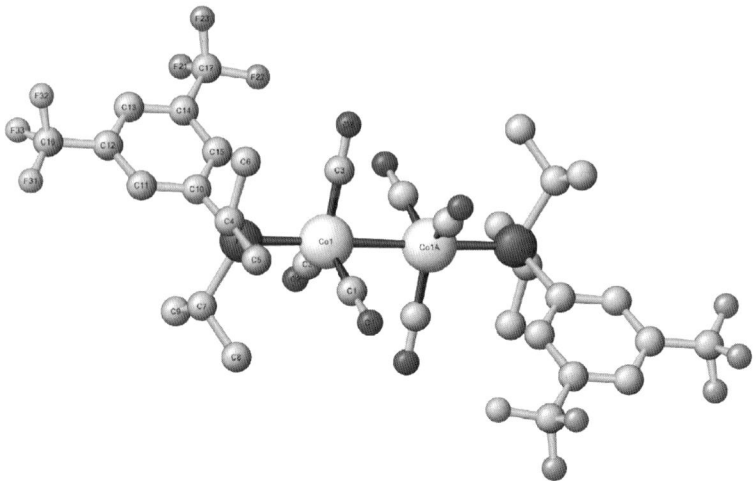

Fig. 10 Molecular structure of $trans$-$Co_2(CO)_6[P(i$-Pr$)_2\{3,5$-$bis(CF_3)$-$C_6H_3\}]$ (**5b**)

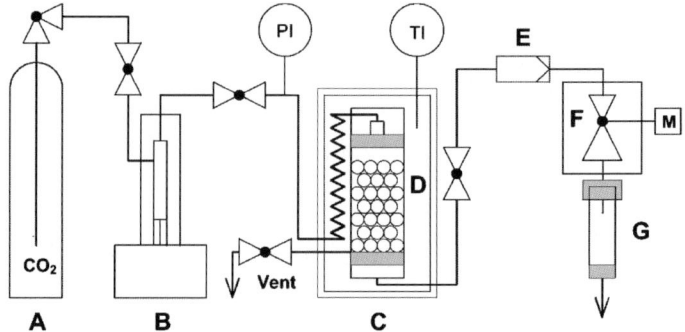

Fig. 11 Extraction device consisting of **A** CO₂ cylinder, **B** syringe pump, **C** oven, **D** extraction cell, **E** particle filter, **F** back-pressure regulator, **G** sampling tube

To allow for solubility measurements by a dynamic procedure, equilibrium conditions have to be established in the extraction cell. If a sufficiently low flow rate is adjusted, the CO_2 passing the extraction cell is loaded with an equilibrium substance amount in the steady state.

The effect of temperature on the solubility of **5b** was investigated in a series of experiments at the same CO_2 density ($\rho = 0.75 \, \mathrm{g \, cm^{-3}}$). The temperature that generally may affect the solubility of volatile compounds in compressed fluids has only a minor impact on the solubility of the relatively low volatile complex **5b** in the investigated range (Fig. 12). At temperatures between 313 and 333 K, approximately the same quantities of **5b** are extracted.

The solubilities of **5b** at 323 K and various pressures are shown in Fig. 13. Dissolved **5b** is not detected below 10 MPa. It is obvious that for these metal complexes having a negligible vapor pressure, the density of scCO₂ is the most important factor influencing the solubility. For **5b**, at 23.7 MPa, the solubility amounts to 16.7 mmol cm⁻³. Compared to the solubilities of different substituted phosphines in scCO₂ published earlier, lower solubilities of the corresponding complexes are expected in general [56, 122, 135].

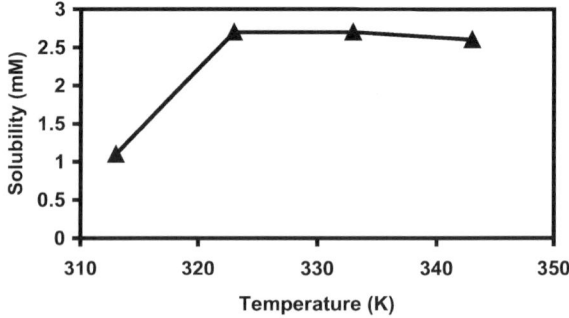

Fig. 12 Solubility of **5b** in CO_2 (ρ: 0.75 g cm⁻³) at various temperatures

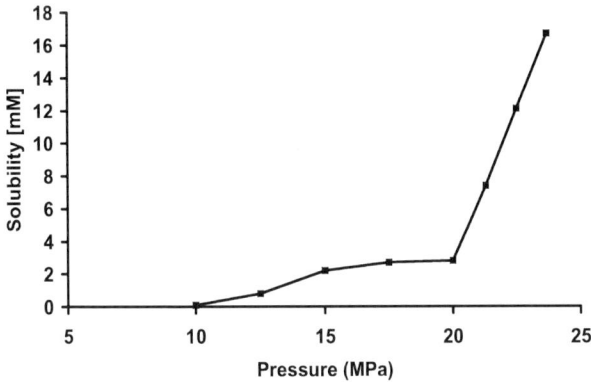

Fig. 13 Solubility of **5b** at 323 K in CO_2 at various pressures

Most recently, Chen et al. studied the solubility of *trans*-$Co_2(CO)_6[P(p$-$CF_3C_6H_4)_3]_2$ in scCO$_2$ ($\rho = 0.45\,\mathrm{g\,cm^{-3}}$) in the presence of 0.74 MPa carbon monoxide in view of utilizing this complex as a pre-catalyst for hydroformylation of ethylene and propylene [123]. The presence of additional $P(p$-$CF_3C_6H_4)_3$ enabled measurements above 373 K without phosphine dissociation against carbon monoxide. Solubilities were measured between $0.2\,\mathrm{mmol\,cm^{-3}}$ (353 K) and $2.1\,\mathrm{mmol\,cm^{-3}}$ (403 K).

For compounds that are usually not available in larger amounts or are expensive, e.g., many transition metal complexes, the static approach was applied as well (Fig. 14). Here, the cell is loaded with a known amount of the solute and scCO$_2$ of known density [124]. This technique is convenient, because it allows for in situ analysis, and the solvatation equilibrium is obtained easily.

The sample is placed into the saturation cell (1) and the system is filled with CO$_2$, using a syringe pump (2) up to the desired pressure. The entire

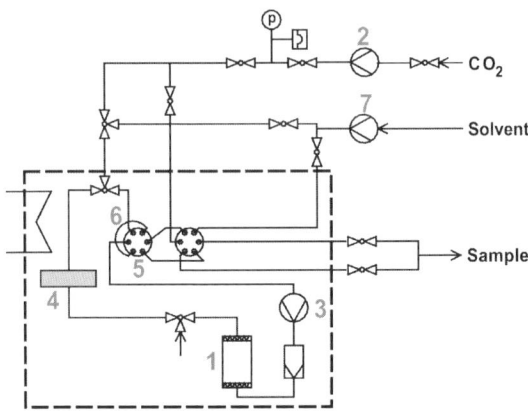

Fig. 14 Static solubility measurement setup

equipment is immersed into a thermostated bath which is kept at the desired temperature. By the circulation pump (3), the CO_2 flow is passed through the saturation cell. When a stationary condition (solution is saturated with the sample) is reached, as detected by in situ UV/VIS measurement (4), the sampling valve (5) is switched and sampling takes place (6). The sample is dissolved in an appropriate organic solvent dosed by an HPLC pump (7) and brought to atmospheric pressure. The concentration of the sample in the organic solvent is determined by means of quantitative UV/VIS spectroscopy. Since the quantity of the solvent rinsing the sample loop and the volume of the sample loop are known, the concentration of the material in $scCO_2$ can be calculated.

Some solid compounds exhibit a significant melting point depression with CO_2 pressure (induced melting) [125]. For example, the melting point of the complex **5b** changes from 383 K (0.1 MPa) to 301 K (6.5 MPa). This phenomenon was taken into account by the development of a saturation cell for liquids and solids that liquefy.

Figure 15 shows the phosphine ligands in *trans*-$Co_2(CO)_6$(phosphine)$_2$ investigated with the static solubility measurement setup. They were synthesized or are commercially available.

A common feature of the solubility curves, as outlined in Fig. 16, is the exponential increase with the pressure due to the increasing density of CO_2. From the solubility measurements [125], it can be concluded that the modification of phosphines by fluorous substituents is not necessarily required to reach sufficient solubilities of the corresponding metal complexes in $scCO_2$. In particular, alkyl substitution on phosphorus promotes complex solubility. Additional aryl groups, e.g., in **1b** and **4b**, cause a reduced solubility of the corresponding complexes. Moreover, *trans*-$Co_2(CO)_6\{(PC_6H_5)_3\}_2$ (**7b**) is completely insoluble in $scCO_2$.

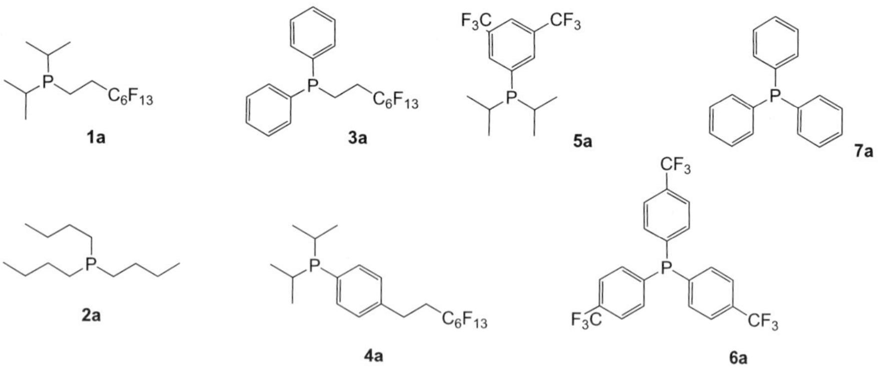

Fig. 15 Phosphines (**1a** to **6a**) used for solubility measurements of the related complexes *trans*-$Co_2(CO)_6$(phosphine)$_2$ (**1b** to **6b**)

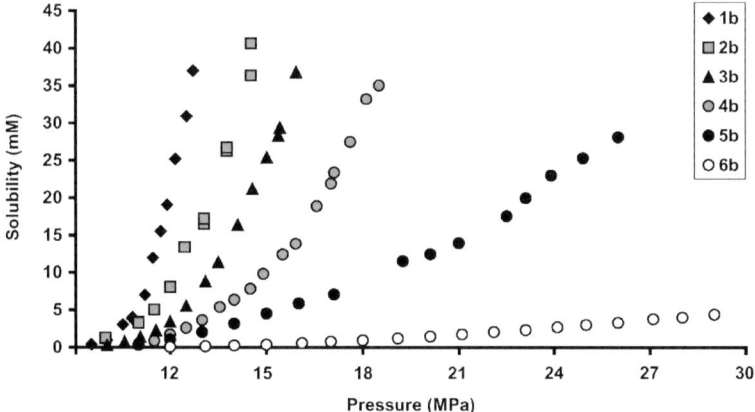

Fig. 16   Solubilities of *trans*-$Co_2(CO)_6$(phosphine)$_2$ in $scCO_2$ at 323 K (ρ: 0.65 to 0.87 g cm^{-3})

It is well known that the presence of fluorous substituents results in a significantly increased solubility of metal complexes in $scCO_2$ (Sect. 3). Consequently, the lower solubility of complexes with aryl-substituted phosphine ligands can be compensated by fluorous substituents, e.g., the complex with $P(p\text{-}CF_3C_6H_4)_3$. The best solubilities are found for complexes with trialkyl phosphine complexes **1b** and **2b**. If one of the substituents is a fluorous alkyl group, the solubility is slightly increased, but the solubility changing effect was found to be minor compared to a P-bonded aryl group substituted by an alkyl group.

Most cobalt complexes investigated in this study are soluble in pure CO_2. Generally, changes in the density (pressure) of $scCO_2$ allow for the enhancement of the solubility. From these data, together with the metal concentration needed for a certain catalyzed reaction, the parameters can be estimated for the use of $scCO_2$ as an appropriate reaction medium, e.g., in the hydroformylation of olefins (Sect. 5). It must be stated that the solubility of such a complex generally is lower compared to the solubility of the corresponding phosphine ligands [55].

3.3
Solubility of $Co_2(CO)_6$(Phosphine)$_2$ in Reaction-relevant Mixtures with Carbon Dioxide

Under the conditions of a catalyzed reaction, a sufficient solubility of a pre-catalyst is the prerequisite for the formation of a catalytically active species in a concentration high enough to ensure optimal homogeneous conditions. Furthermore, the solubility-changing influences of the reacting components must be considered. Therefore, solubility measurements were performed to investigate the influence of substrate and product at concentrations typical of a hydroformylation as a prototypical catalytic reaction. The applied conditions were chosen such that no chemical interaction with the complexes takes place. This was verified by in situ UV/VIS measurements that did not reveal any changes in the absorption pattern of the complex.

It is well known that the solubility of a compound in $scCO_2$ may be increased by adding modifiers (Sect. 2.3). One would expect an increased solubility of complexes in $scCO_2$ in the presence of an olefin/aldehyde, in which these complexes are well soluble at atmospheric pressure. Surprisingly, the example of *trans*-$Co_2(CO)_6[P(n$-$Bu_3])_2$ (**2b**) shows that the solubility is lower in the presence of olefins compared to that solubility in pure CO_2. For concentration ranges typical of a catalyzed reaction, the solubility of a metal complex drops with increasing chain length of an added α-olefin (Fig. 17)

An analogous fluorinated complex derivative measured under the same conditions also is less soluble and exhibits similar pressure differences of the initial solubility with and without olefin addition (Fig. 18). This gives rise to the assumption that the observed "negative modifier effect" generally is valid also for other derivatives.

The additional presence of 1-nonanal serves as model for the formation of the linear hydroformylation product of 1-octene and its influence on catalyst solubility. The aldehyde lowers the complex solubility further (Fig. 19) as compared to the mixture with olefin only.

Up to now, the solubility decrease of cobalt complexes with these modifiers has not been explained satisfyingly. It is assumed that the changes in the solvatization characteristics observed are caused by different interactions of the solute with the mixture of organic components and CO_2; the modifier-solute (olefin/aldehyde–complex) interaction probably is stronger than the solute–$scCO_2$ interaction. Future theoretical treatment may also improve the

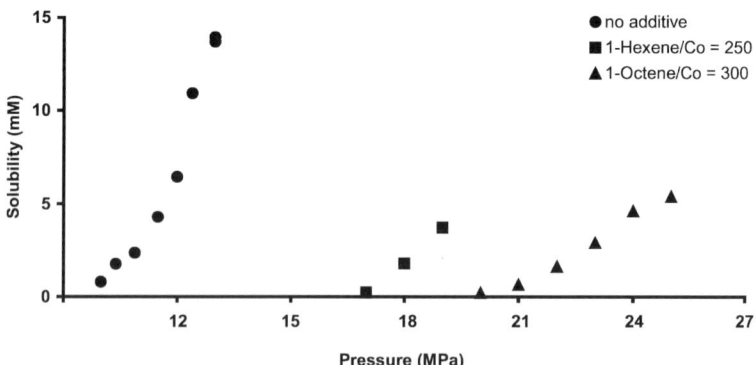

Fig. 17 Solubility of *trans*-$Co_2(CO)_6\{P(n$-$Bu_3)_3\}_2$ (**2b**) in pure $scCO_2$ and with added α-olefins at 323 K (ρ: 0.65 to 0.85 g cm^{-3})

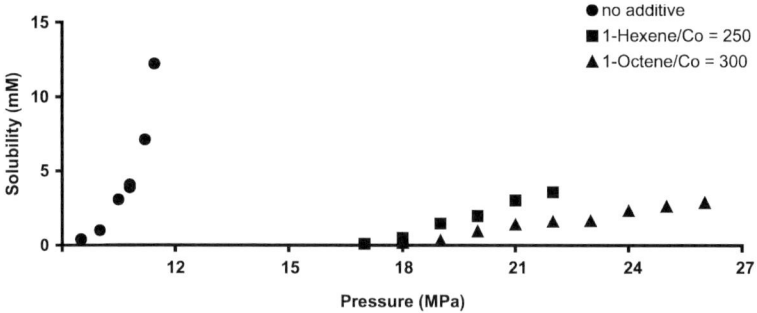

Fig. 18 Solubility of *trans*-Co$_2$(CO)$_6${P(i-Pr)$_3$)$_2$C$_2$H$_4$C$_6$F$_{13}$}$_2$ (**2a**) in pure scCO$_2$ and with added α-olefins (ρ: 0.65 to 0.85 g cm^{-3})

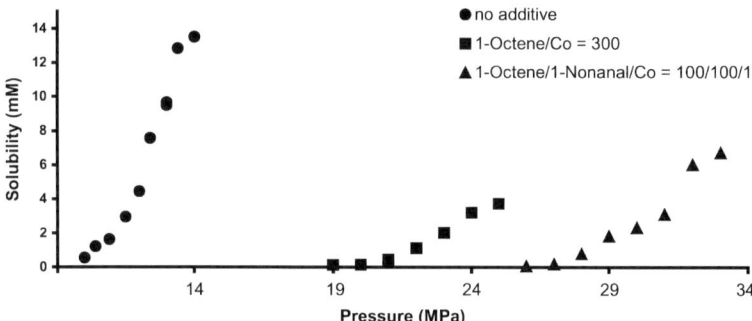

Fig. 19 Solubility of *trans*-Co$_2$(CO)$_6${P(n-Bu$_3$)$_3$}$_2$ (**2b**) in pure scCO$_2$, with added 1-octene and with 1-octene/1-nonanal at 323 K (ρ: 0.65 to 0.91 g cm^{-3})

understanding of the different types of interaction (e.g., clustering) at those complexes.

Those solvatization effects that reflect the ongoing conversion of the hydroformylation are supposed to be similar for catalytically active complexes with molecular similarity. This knowledge opens up the perspective of a pressure-induced catalyst separation in a catalyzed reaction in a desired conversion or a desired operation range of the reactor pressure, where no dissolved metal complex remains in the CO$_2$ phase a defined point (Sect. 5.5).

4
Multi-Phase Approaches
to Homogeneously Catalyzed Reactions Utilizing scCO$_2$

The use of scCO$_2$ either alone [135–137] or in multi-phase systems [138–140] offers interesting process alternatives to other methods of catalyst immobilization such as heterogenization on solid supports [126–128], aqueous biphasic systems [129], ionic liquids [130–132] or fluorous systems [133, 134]. Table 1 gives an overview and compares several of these options [141].

For instance, catalysis in liquid/liquid two phases is generally referred to as biphasic catalysis and has widened the practical scope of homogeneous catalysis; the catalyst is present in one liquid phase, while reactants and products are present in the other liquid phase. Thus, the catalyst can be separated by simple phase separation. Celanese is operating a 300 000 t/a plant for propylene hydroformylation using a water-soluble rhodium phosphine complex in a biphasic mode of operation at the Ruhrchemie site in Oberhausen [142].

These multi-phase approaches, however, suffer from the drawback that some of the reaction media must be separated from the catalyst or the reaction products. Therefore, the conditions of catalyst separation may be very different from those of the reaction such that catalyst decomposition may still occur. Furthermore, most approaches have two or three phases during the reaction, which may cause problems in controlling the phase equilibria and in controlling the distribution of the reacting components and the catalytically active component between the phases. Most effective agitation of the reaction mixture is often required.

One of the benefits of scCO$_2$ for homogeneous catalysis is that rates or selectivities may be significantly higher than in other multi-phase systems or in conventional solvents, because mass transfer across interfaces is enhanced. An example is CO$_2$ hydrogenation that simultaneously uses CO$_2$ as both reaction medium and substrate [114].

Separation, controllable through changes in pressure and/or temperature and/or composition (with respect to increasing substrate conversion), may take place in two ways, as outlined in Fig. 20. Both ways, A and B represent

Table 1 Multi-phase catalysis approaches with respect to both, reaction and separation conditions (S: substrates, P: products)

Type	Phase A	Phase B	Separation Principle	Phase A	Phase B
Aqueous Two-Phase	H$_2$O (cat.)	organic (gas)	Enrichment of products in the organic phase	H$_2$O (cat.)	organic (P)
Fluorous Two-Phase	RF-solvent (cat./S)	(gas)	Product separation through change of T	RF-solvent (cat.)	organic (P)
IL Two-Phase	IL (cat.)	organic (gas)	Enrichment of products in the organic phase	IL (cat.)	organic (P)
IL/CO$_2$ Two-Phase	IL (cat.)	organic (gas)	Product extraction by CO$_2$	IL (cat.)	organic (P)
scCO$_2$	CO$_2$ (cat. / S)	–	Product separation through change of density	CO$_2$ (cat.)	organic (P)
scCO$_2$/H$_2$O Two-Phase	CO$_2$ (cat.)	H$_2$O/S	Enrichment of products in the aqueous phase	CO$_2$ (cat.)	H$_2$O (P)

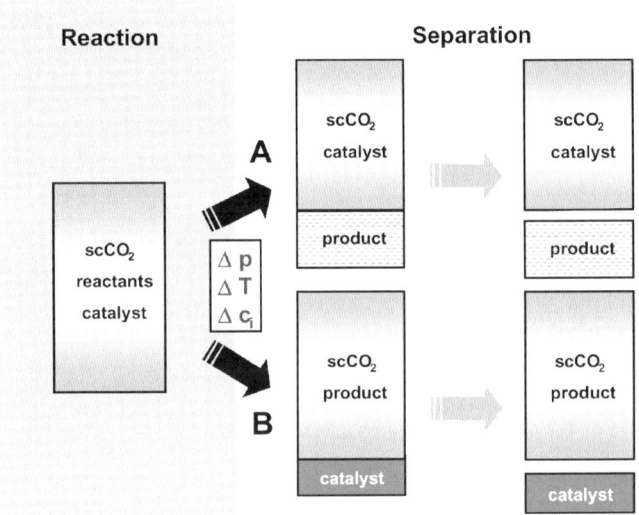

Fig. 20 Reaction/separation strategies for multi-phase catalysis in $scCO_2$

elegant approaches to an integrated product/catalyst separation, because the interaction of these control parameters can be adjusted to the requirements for each process. For example, thermosensitive catalysts could be made insoluble and, hence, separable by reducing the pressure for the separation step without the need to separate the solvent thermally.

An ideal system would enable continuous operation with separation under conditions similar to those of the reaction itself, as a result of which all the catalyst remains in its active state for all times. Use of $scCO_2$ as a reaction medium would not only allow for reaction and separation under similar conditions to reduce the problems of catalyst deactivation, but would also have a considerable effect on the plant design. Its total lack of flammability, contrary to conventional organic solvents, reduces safety problems and makes $scCO_2$ attractive for homogeneously catalyzed reactions, particularly for oxidation and epoxidation reactions [143].

In the past, the majority of "high-pressure" homogeneous catalytic reactions were conducted in batch systems, which may cause problems in scale-up for SCFs because of the higher pressures needed for achieving the supercritical state. Therefore, continuous processing has also been investigated in the last years. It would be preferable for industrial-scale SCF reactions, because it involves smaller and, hence, safer equipment [144–150]. In addition, capital costs are likely to be lower than in batch systems.

5
Hydroformylation Processes Utilizing Carbon Dioxide

5.1
Motivation for Using Carbon Dioxide
as an Alternative Reaction Medium in Hydroformylation

The hydroformylation of olefins discovered by Otto Roelen [151] is one of the most important industrial homogeneously catalyzed reactions [152, 153] for the synthesis of aldehydes with an estimated production of more than 9.2 million t in 1998 [153]. Hydroformylation is the addition of hydrogen and carbon monoxide to a C,C double bond. Industrial processes are based on cobalt or rhodium catalysts according to Eq. 1. The desired products are linear (*n*-) and branched (*i*-) aldehydes, in which the linear products are generally favored for subsequent processing.

Equation 1

The catalytic cycle proposed by Heck and Breslow [154] consists of a series of elementary steps where, depending on the catalyst and reaction conditions, the rate-limiting step is usually associated with the oxidative addition of H_2. Technologically speaking, hydroformylation has the drawbacks of a complicated sequence of chemical and/or physical operation units and of a complex product and catalyst separation. Consequently, high consumptions of materials are the results. Industrial hydroformylation is a homogeneously catalyzed reaction of gaseous or liquid olefins and gaseous syngas (CO/H_2). It is known that one major difficulty is the efficient contacting of the reactants and the catalyst [22, 153, 175–177]. The influence of transport phenomena on the overall reaction rate has to be taken into account, especially at low pressures [153]. The mass transfer and separation problems are increased by reacting long-chained olefins. They are currently not used as substrates on a technical scale for rhodium catalysis due to the limited thermal stability of the rhodium catalysts [178] as well as for the hydroformylation in an aqueous reaction medium due to the limited solubility of the higher olefins [152, 153, 179]. Therefore, efforts have been undertaken to find new possibilities of intensifying the process.

Different approaches to overcome the discussed limitations have been published [152, 153, 177]. Wiese et al. reported that space-time yields increased by a factor of ten when using a packed tubular reactor and altered operating conditions compared to those in a conventional stirred tank reactor [176].

For the rhodium-catalyzed hydroformylation of propylene in an aqueous biphasic system, Cents et al. have shown that the accurate knowledge of the mass transfer parameters in the gas–liquid–liquid system is necessary to predict and optimize the production rate [180]. Choudhari et al. enhanced the reaction rate by a factor of 10–50 by using promoter ligands for the hydroformylation of 1-octene in a biphasic aqueous system [175].

For this reason, $scCO_2$ was investigated as solvent substitute in the hydroformylation process.

Since the pioneering work by Rathke's group [155, 156], several research groups have investigated the hydroformylation in SCFs, particularly in $scCO_2$ [88, 135, 136, 157–174]: (i) The use of $scCO_2$ as an alternative reaction medium is one of several approaches that have been developed to intensify the process by using its beneficial properties, e.g., its solvent characteristics and the gas-like diffusion properties, for eliminating the mass transfer resistances specific to the gas-liquid reactions to process the hydroformylation in a homogeneous phase. This also includes heterogeneously catalyzed hydroformylation approaches; (ii) a second approach is to valorize the potential of CO_2 for a simplified separation of the products and the catalyst by a simple depressurization or temperature change when CO_2 changes from the supercritical to the gaseous state; (iii) CO_2 can be used as reactant, too, and could serve as CO source in the hydroformylation.

In view of their importance to process improvement, fundamental aspects of and progress achieved by using $scCO_2$ in hydroformylation as well as experimental techniques shall be described in the following sections.

5.2
Homogeneously Catalyzed Hydroformylation in scCO$_2$

In the initial study of propylene hydroformylation in $scCO_2$ with unmodified and modified cobalt catalysts, equilibrium and dynamic processes were investigated by means of high-pressure NMR spectroscopy [155, 156]. One aspect of this work was to capitalize on the gas miscibility and viscosity reduction in $scCO_2$ for homogeneous organometallic catalysis, which had been studied previously for very different reaction types only [180–189]. At 353 K in $scCO_2$ (ρ_{CO_2} : 0.5 g cm^{-3}) an aldehyde formation rate of 0.77×10^{-5} mol s^{-1} was determined, similar to the value of 1.2×10^{-5} mol s^{-1} determined for 1-octene in methylcyclohexane as solvent under comparable conditions. The same group also studied modified cobalt catalysts with high-pressure NMR spectroscopy [123]. Studies for comparing the activities of $Co_2(CO)_8$ in the hydroformylation of 1-octene in $scCO_2$ and toluene under similar conditions were reported recently [174]. The decreasing pressure as a function of the consumption of syngas with the reaction progress is presented in Fig. 21 (left). The results reveal that the activity of $Co_2(CO)_8$ in $scCO_2$ is similar to that in toluene. Analogous conclusions were reported earlier for the hydroformylation of 1-hexene using rhodium triethylphosphine catalysts [168] and rhodium catalysts with fluorous phosphines [136, 190]. In the hydroformylation of 1-octene under the conditions presented in both reaction media, toluene and $scCO_2$, n-nonanal (linear aldehyde), i-nonanal (2-methyloctanal as branched aldehyde) as well as cis- and $trans$-2-octene were formed. Hydrogenation products, octane, 1-nonanol, and 2-methyl-octanol are further minor byproducts. The conversion and selectivities after 20 h in both toluene and $scCO_2$ are presented in Fig. 21 (right).

Similar reaction rates in toluene and $scCO_2$ were reported by Cole-Hamilton et al. for the hydroformylation of 1-hexene on a catalyst prepared in situ from $[Rh_2(OAc)_4]$ and PEt_3 with a slightly higher n/i-selectivity in $scCO_2$ [168]. Guo et al. [158] reported an increased n-butanal selectivity of 88% in $scCO_2$ in comparison with 83% observed in benzene [191]. In the hydroformylation of propylene, the n/i ratio was slightly influenced by pressure and temperature. At a constant temperature of 361 K, the selectivity of n-butanal increases from 73 to 81% by raising the CO_2 pressure from 0.89 to 1.78 MPa [158]. A significant rate enhancement was reported by Leitner's group for the rhodium-catalyzed hydroformylation of various olefins (1-octene, $trans$-3-hexene, styrene, allyl acetate) in $scCO_2$ and, for comparison, in toluene by using $[(cod)Rh(hfacac)]$ as catalyst precursor (Eq. 2) [136].

After 20 h, conversion was found to be higher in $scCO_2$ than in toluene under similar reaction conditions (Table 2) [136]. This effect is more pronounced for internal olefins, such as $trans$-3-hexene, compared to terminal olefins, such as 1-octene.

Another example of a reaction rate increased by a factor of 10 to 20 through the replacement of the organic solvent by $scCO_2$ was reported by Xiao et al. [193]. Whereas the hydroformylation of methyl, butyl, and

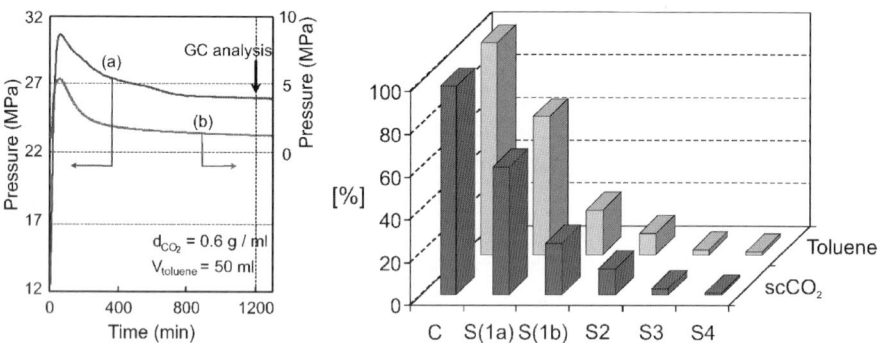

Fig. 21 Hydroformylation of 1-octene with $Co_2(CO)_8$ in $scCO_2$ (**a**) and in toluene (**b**); initial conditions: 53 mmol of 1-octene, 106 mmol syngas ($H_2 : CO = 1 : 1$), 0.106 mmol $Co_2(CO)_8$, T = 393 K; *left*: Reactor pressure as function of time; *right*: C: Conversion, S: Selectivity, 1a: 1-nonanal, 1b: 2-nonanal, 2: $\sum$ nonanols, 3: $\sum$ octenes (without 1-octene), 4: n-octane

$$R\diagup\diagdown R \xrightarrow[\substack{scCO_2,\ 313\text{-}338\ K \\ 2\text{-}6\ MPa\ CO/H_2\ (1:1)}]{\substack{CO/H_2 \\ [Rh(hfacac)(cod)]\ /\ PR_3}} OHC\diagdown\diagup R\ (R) + R\diagdown\diagup R\ (CHO)$$

Equation 2

Table 2 Hydroformylation of various olefins in $scCO_2$ and toluene using $[(cod)Rh(hfacac)]$ as catalyst precursor [a]

Olefin	$\rho\ (g\,cm^{-3})$	T (K)	Conversion [c]	Aldehyde distribution (%)
1-octene	0.55	313	> 97	n-nonanal (57), i-nonanal (39), 2-propylhexanal (3), 3-ethylheptanal (1)
1-octene	[d]	313	54	n-nonanal (58), i-nonanal (42)
trans-3-hexene	0.54	313	> 97	2-methylhexanal (14), 2-ethylpentanal (86)
trans-3-hexene	[d]	313	23	2-methylhexanal (4), 2-ethylpentanal (96)
2,3-dimethyl-2-butene	0.55	318	0	
styrene	0.51	333 [b]	> 97	2-phenylpropanal (85), 3-phenylpropanal (15)
allyl acetate	0.58	333 [b]	> 97	4-oxobutyl acetate (28), 3-oxo-2-methylpropyl acetate (72)

[a] Conditions: Substrate (4.0–6.0 mmol), $[(cod)Rh(hfacac)]$ (5.7–6.7 μmol), $V_{reactor}$: 25 cm^3, p(CO/H$_2$): 4.44 MPa at room temperature, t: 20 h,
[b] p(CO/H$_2$): 3.95 MPa at room temperature,
[c] Total conversion to aldehydes: %,
[d] Control experiment in toluene (25 cm^3, $V_{reactor}$: 50 cm^3)

t-butyl acrylate on Rh-P(p-C$_6$H$_4$C$_6$F$_{13}$)$_3$ in toluene yields TOF values less than 200 h^{-1}, up to more than 1,800 h^{-1} were determined in scCO$_2$ under the same conditions (Table 3). The significant enhancement in hydroformylation rates for acrylates in scCO$_2$ was caused by the solvent–solute interactions and by shifting the equilibrium in favor of the unsaturated key intermediate as a result of a carbonyl-CO$_2$ donor–acceptor interaction, as shown in Eq. 3.

The effects of phosphine ligands, CO$_2$, and syngas pressure on the hydroformylation of 1-hexene were investigated by Arai's group [190]. The nature of the phosphine ligand does not change considerably the n/i selectivity. Interestingly, it was found that the aldehyde yield goes through a minimum at

Table 3 Hydroformylation of acrylates by Rh-P(p-C$_6$H$_4$C$_6$F$_{13}$)$_3$ in scCO$_2$ and toluene [a]

Substrate	Solvent	[Olefin] [b]	Conversion [c]	Aldehydes [%] [d]	TOF, h^{-1} [e]
Methyl acrylate	scCO2	0.45	40.8	97.6	1593
Butyl acrylate	scCO2	0.28	42.9	97.5	1671
^{t}Butyl acrylate	scCO2	0.28	46.9	97.4	1827
Methyl acrylate	Toluene	0.45	2.3	99.6	92
Butyl acrylate	Toluene	0.28	2.1	99.8	84
^{t}Butyl acrylate	Toluene	0.28	4.8	99.6	191

[a] Reactions were carried out at P(p-C$_6$H$_4$C$_6$F$_{13}$)$_3$/[Rh(acac)(CO)$_2$] = 10, olefin/rhodium = 4000, T: 353 K, P_{CO_2}: 17.7 MPa CO$_2$, and P_{syngas}: 1.97 MPa (H$_2$/CO = 1) for 1 h
[b] Olefin concentration: mol L^{-1}
[c] Conversion of the acrylates: %
[d] Selectivity for branched aldehydes. The linear aldehydes were not detected by the GC. The hydrogenated product of propionates accounts for the product balance
[e] Average turnover frequency

Equation 3

about 9 MPa with increasing CO$_2$ pressure and tends to increase with increasing syngas pressure.

Recent studies [174] also focused on the influence of the Co/phosphine ratio on the catalytic performance of *trans*-Co$_2$(CO)$_6$[P(3 – FC$_6$H$_4$)$_3$]$_2$ **8** with additional P(3 – FC$_6$H$_4$)$_3$ in the hydroformylation of 1-octene in scCO$_2$. The results are summarized in Fig. 22.

When using Co$_2$(CO)$_8$ (P/Co = 0), conversion of higher than 95% was achieved. Replacing two CO ligand groups in Co$_2$(CO)$_8$ by tris(3-fluorophenyl)phosphine and use of this pre-catalyst designated **8** (P/Co = 1) decreases the conversion to 92%, with the rate further decreasing with excessing phosphine. Compared to the unmodified Co$_2$(CO)$_8$, **8** provides for an enhanced aldehyde selectivity from 76–92%, accompanied by an increased *n/i* ratio from 2.3 to 3.3. The share of branched aldehyde remains constant and is practically independent of the excess of phosphine. These findings are unexpected, because a large number of both scientific and technical reports [179, 180] indicated a high excess of phosphine to be a necessary prerequisite for effectively increasing the aldehyde selectivity. Nevertheless, finding of a cata-

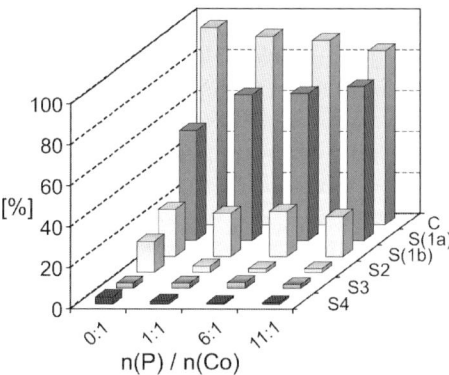

Fig. 22 Conversion and product distribution by the hydroformylation of 1-octene in scCO$_2$ with Co$_2$(CO)$_8$ and **8** with different P : Co atomic ratios. Initial conditions: 53 mmol of 1-octene, 106 mmol syngas (H$_2$/CO = 1), 0.106 mmol Co$_2$(CO)$_8$, T: 393 K. C: Conversion, S: selectivity, 1a: 1-nonanal, 1b: 2-nonanal, 2: $\sum$ nonanols, 3: $\sum$ octenes (without 1-octene), 4: *n*-octane

lyst that works preferably without an excess of expensive ligand is aimed for technical application. Likewise, hydrogenation was substantially suppressed from 16.2 to 3.2% for alcohols and from 4.6 to 1.6% for alkane.

Examples of a negative influence of the use of scCO$_2$ are also known [194]. This shows that the phenomenology of such reactions in scCO$_2$ is not yet well understood and that supplementary research data are needed.

5.3
Heterogeneously Catalyzed Hydroformylation in CO$_2$

In many cases, gas-liquid-catalytic processes exhibit mass transfer limitations. By using solid catalysts, supplementary resistances are involved [8c]. In a three-phase reaction, the following physical and chemical steps are possible: (a) Transfer of gaseous reactant from the bulk gas phase to the gas/liquid interface (diffusion) and (b) from the interface to the bulk liquid phase (absorption and diffusion); (c) transfer of both reactants (gas and liquid) from the bulk liquid to the external surface of the catalyst particle (diffusion through a stagnant external film surrounding the catalyst particle); (d) transfer of reactants into porous catalyst (internal diffusion); (e) adsorption of reactants following either step (c) or (d); (f) surface reaction; (g) desorption and transfer of product(s) by (h) internal and (i) external diffusion to the bulk liquid or gas phase. All steps may be accompanied by heat transfer. The process rate is determined by the slowest step. Under supercritical conditions, all reactants are present in the single-fluid phase, which is contacted with the solid catalyst. The gas/liquid transfer resistance is eliminated (steps a and b) and the external fluid film diffusion resistance (step c) is lowered due to the lower viscosity of SCFs. Not only the process rate but also product selectivity and/or separation may be simplified by an appropriate selection of the reaction medium and processing parameters [182]. The process phenomenology in heterogeneous catalysis, combined with the potential of SCFs, is the subject of several reviews and monographs [20, 195–198]. Particular aspects of heterogeneously catalyzed reactions in SCFs are presented in several reports [170, 172, 173, 183, 184].

Both heterogeneous and heterogenized hydroformylation in scCO$_2$ shall be discussed in the following paragraphs, because the technological tasks are similar. Since heterogeneous catalysts are defined as active catalysts supported on solids and generally characterized by their porosity, heterogenized catalysts include a wide variety of anchored or immobilized homogeneous catalysts on different supports such as organic polymers and inorganic supports or combinations of these [152].

Rhodium, supported on activated carbon, is active for the hydroformylation of propylene to butanal, but with a n/i selectivity below 1.5 [170]. The determined activation energy of 8.3 ± 2.2 kJ mol^{-1} indicates that the reaction is controlled by the rate of mass transfer. The catalysts supported on activated carbon exhibit an irreversible adsorption of propylene, with negative consequences for the catalyst performances. By using appropriate precursor species during preparation, the surface functionality of the support can be modified over a large range from a hydrophobic to a hydrophilic or to a CO$_2$-philic one, corresponding to the desired target [172, 173]. On a hydrophilic modified support, the irreversible adsorption of olefins is inhibited, but the reaction rate is low. A catalyst obtained by the deposition of rhodium on silica gel with a CO$_2$-philic modified surface reaches better performances in hydroformylation. The local concentration of CO$_2$ around the rhodium-active center is assumed to be higher. Consequently, the irreversible adsorption of olefins is suppressed and the transport of both reactants to the active center and generated products from the active center to the bulk fluid phase is facilitated.

Metal leaching remains a serious problem, as it leads to a loss of activity. Reduced leaching has been observed when the catalysts are anchored inside the pores of zeolites (so-called *ship-in-a-bottle* catalysts) or on mesoporous solids [140, 173]. A very rare example of leaching being reduced to an acceptable level was reported by Sandee et al. [171, 199]. For the catalyst preparation, a modified XANTPHOS ligand with a hydrocarbon chain terminated by a triethoxysilyl group was used. Incorporation in a sol-gel solution yields a modified silica support with the anchored ligand on the surface, and the rhodium is effectively bound on this support. This catalyst has been used successfully for the continuous hydroformylation of 1-octene [169]. Despite their simple separation, heterogeneous catalysts are generally characterized by an incomplete utilization of the active metal and, consequently, by a lower activity compared to analogous homogeneous catalysts.

5.4
Hydroformylation in Ionic Liquids/scCO$_2$

Ionic liquids are low-melting organic salts that are not volatile and not soluble in scCO$_2$. However, scCO$_2$ dissolves in ionic liquids up to high concentrations, resulting in so-called expanded ionic liquids [200–204]. The dissolved scCO$_2$ can also be used for extracting organic compounds from ionic liquids [204–206]. These special properties of both ionic liquids and scCO$_2$ were combined for hydroformylation [204]. An ionic catalyst is immobilized in the ionic liquid in a stirred tank reactor. The syngas and the olefin as well as the scCO$_2$ transport medium are passed into the reactor either separately or mixed. The reaction takes place under homogeneous conditions. The product flow dissolved in scCO$_2$ is then subjected to product separation from the gases by simple expanding. The gaseous CO$_2$ containing CO and H$_2$ can be recompressed and recycled. The catalyst and the ionic liquid remain in the reactor. This concept was investigated for the continuous hydroformylation of long-chained olefins. The performances of this alternative process are similar to those of various industrial processes.

5.5
Carbon Dioxide as a Switchable Transport Vector for Simplified Separation

As described in previous sections, scCO$_2$ is able to dissolve simultaneously gaseous and liquid or solid reactants with a liquid or solid catalyst. The dissolution is promoted by an increased temperature, CO$_2$ pressure or CO$_2$ density [56, 117, 207]. Thus, it is possible to convert a multi-phase system corresponding to the separation step into a homogeneous one during reaction. By switching temperature or pressure after the reaction, the CO$_2$ is transferred into its gaseous state and may simply be recycled. Subtle control of this operation allows for the selective separation of products or catalyst together with CO$_2$ [136, 174]. This separation principle by combination of catalysis and extraction using supercritical solutions (CESS process) was first demonstrated by Leitner et al. and later for the rhodium-catalyzed enantioselective hydroformylation of prochiral olefins [208].

Since alkylphosphine-modified rhodium or cobalt catalysts are moderately soluble in scCO$_2$, their arylphosphine analogs are often insoluble or soluble only at higher pressure and sometimes at higher temperature [117]. Recent studies particularly showed this behavior of catalysts that were insoluble or less soluble under subcritical conditions and soluble in the supercritical state. They were applied successfully for an elegant separation of the products from the catalyst to achieve a catalyst recycling [174]. It was found that 8 has suitable properties, such as a strong dependence of its solubility on CO$_2$ density, for its simple separation after the reaction. This allows for a mechanical separation of the catalyst by temperature and pressure swings. Batch experiments were performed in the experimental setup presented in Sect. 5.7 (Fig. 24), in which two parallel reactors were used. After 20 h the reaction was stopped, and the products, together with the unconverted syngas and CO_2, were expanded with a fine back-pressure regulator to a gas-liquid separator. The separated catalyst was recycled by six runs *a...f* in the first reactor and *a'...f'* in the second reactor, respectively. The fresh charges of the reactors and the reaction parameters were identical in runs *a* and *a'*, respectively. In addition to the regenerated catalyst, additional tris(3-fluorophenyl)phosphine was added to reactor R1 prior to the runs *c*, *e*, and *e'*. The results are given in Fig. 23 (A and B). Using a stoichiometric amount of syngas in relation to 1-octene (Fig. 23A), the conversion amounted to 96–99% (runs *a* and *b*). The aldehyde yield decreased from 88% (run *a*) to 78% (run *b*). A significant increase from 3.5% to 12% was determined for olefin isomerization. Addition of phosphine prior to the run *c* led to a decrease of the isomer formation and, thus, improved the *n/i* selectivity. However, excess phosphine resulted in a slower conversion rate without improved aldehyde selectivity. Both conversion and aldehyde selectivity increased slightly from run *c* to run *d*. This phenomenon may be explained by the partial removal of the phosphine when discharging CO_2 and the products from the reactor after run *c*. Due to the partial loss of volatile phosphine, further phosphine needs to be added to the remaining catalyst to support a complete catalyst reformation. The second addition of phosphine prior to the run *e* extended the catalyst lifetime. After six successive hydroformylation runs with **8**, aldehyde conversion and selectivity were still above 60 and 70%, respectively.

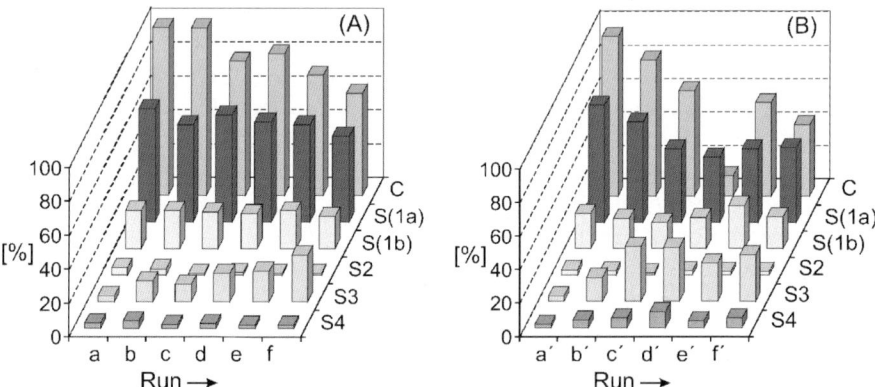

Fig. 23 Conversion and product distribution by the hydroformylation of 1-octene in scCO$_2$ with recycled catalyst 8. Initial conditions **a**: 53 mmol of 1-octene, 106 mmol syngas ($H_2 : CO = 1 : 1$), 0.106 mmol $Co_2(CO)_8$, T = 393 K; **b**: 53 mmol of 1-octene, 159 mmol syngas ($H_2 : CO = 1 : 1$), 0.106 mmol $Co_2(CO)_8$, T = 393 K. C: Conversion, S: selectivity, 1a: 1-nonanal, 1b: 2-nonanal, 2: $\sum$ nonanols, 3: $\sum$ octenes (without 1-octene), 4: *n*-octane

These results promote a potential use of cobalt catalysts instead of expensive rhodium-based catalysts for the hydroformylation in $scCO_2$ [16, 17, 209, 210]. Although the main goal of this study was to demonstrate catalyst recycling, it must be noted that the catalytic performances have been obtained at process parameters and under conditions that were far from being optimized.

High CO partial pressure promotes the formation of unmodified cobalt catalyst [211, 212]. Using a 50% excess of syngas (Fig. 23B), a reduced activity was observed. The conversion decreased from 96% (run a') to 12.2% (run d') after four runs. In addition, chemoselectivity decreased considerably. Both hydrogenation, particularly to octane, and isomerization increased.

5.6
Carbon Dioxide as a Carbon Monoxide Source

The hydroformylation of alkenes to aldehydes or directly to alcohols by using CO_2 as CO source (Eq. 4) was reported by Tominaga and Sasaki [214, 215].

Equation 4

The reaction proceeds in three steps: (i) CO_2 conversion to CO; (ii) hydroformylation to the aldehyde and (iii) hydrogenation to the alcohol. In addition, conventional solvents (N-methylpyrrolidone, 1,3-dimethylimidazolidinone, dimethoxyethane, toluene, benzene, tetrahydrofurane) were used to dissolve the additives, e.g., LiCl, LiBr, and LiI, which were found to be crucial for the reaction. Conversions of up to > 99% and high alcohol selectivities of up to 91% for cyclohexene as substrate were reported. Compared to the hydroformylation using CO directly under similar conditions, the use of CO_2 produces higher yields for different olefins. The basic advantage of this concept is the finding of a new alternative for the activation of less reactive CO_2, combined with an olefin refinement. Due to its abundance and to reach a sustainable process management, the recycling of CO_2 is an exciting as well as a necessary research area.

5.7
Technique and Apparatus for Investigations of Reactions in $scCO_2$

The wide diversity and the large number of apparatuses used for investigations and processing of reactions in SCFs make a systematization difficult. They have been developed during the time from the first "digester" by Denys Papin in 1680 to Ipatiev's high-pressure "bomb" [198] and up to the first industrial plant by Thomas Swan & Co. Ltd. developed as a result of the successful collaboration with a research group from the University of Nottingham [213]. Numerous examples of apparatuses and methods used for investigations of catalytic reactions in $scCO_2$ have been reported in the literature. A quite extensive (but so far not complete overview) is offered in [198] and the citations therein. The setups used for the investigation of catalytic reactions in $scCO_2$ in our laboratories are presented in the following [174].

As discussed in previous chapters, the phase behavior with changing temperature and pressure may be strongly influenced by small concentration gradients in multi-component systems already. Therefore, experimental control should take this into account. It is a common practice to use reactors with glass or sapphire windows. The transition of an inhomogeneous multi-phase system to a homogeneous one can be observed visually as cloud point (Sect. 2.2, with the pressure and temperature values being monitored.

Generally, such investigations in SCFs are time consuming. The number of variable parameters makes the investigations with a single reactor vessel troublesome. In contrast, multi-reactor systems allow to benefit from both the evolution of theoretical combinatorial chemistry and high-throughput experimentation methods. The advantages of these systems include a higher productivity, wider coverage of the parameter space, and, therefore, a bet-

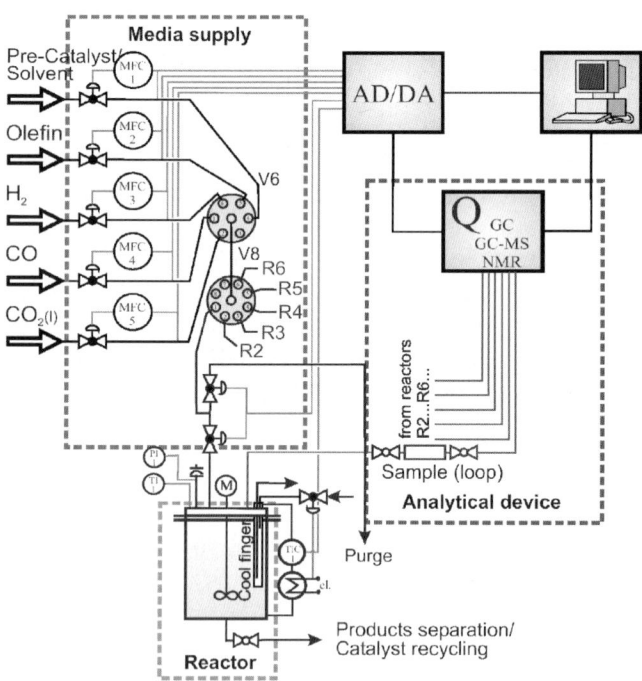

Fig. 24 Multi-reactor system for parallel screening of catalytic reactions in $scCO_2$

ter explanation of the effects observed. A multi-purpose multi-reactor system for parallel screening of catalytic reactions has been developed at the Institute for Technical Chemistry of Forschungszentrum Karlsruhe (Fig. 24). The system consists of: (i) A parallel reaction unit; (ii) a fully automated media supply unit, and (iii) an analytical device unit. The reaction unit includes six $110 \, cm^3$ stainless-steel vessels (R1 to R6) operating up to 35 MPa and 623 K, each one equipped with inlet and exhaust valves, a safety rupture disk, and a pressure transducer in addition to an internal K-type thermocouple. A PID temperature controller, electrically heated jackets, and internal cooling fingers serve to control the desired temperature with an accuracy of ± 1 K. Agitation takes place by a magnetically coupled stirrer. The temperature, pressure, and stirring speed are recorded continuously by a PC, thus ensuring online monitoring and evaluation of the reaction progress. The media supply unit includes an HPLC pump for dosing liquid compounds and solutions, two high-pressure mass flow controllers for CO and H_2, and a syringe pump for dosing the liquid CO_2. Solid compounds are introduced via stainless-steel tubes integrated in the inlet valves. The analytical device unit consists of a sampling loop system and a GC system equipped with an FID detector for organic compounds as well as an TCD detector allowing for quantitative gas analyses, e.g., CO and H_2.

The system can be operated in the parallel mode, discontinuously (batchwise) with each reactor as an independent unit, semi-continuously or as a reactor cascade. Both homogeneous and heterogeneous reactions as well as product and catalyst separation and catalyst recycling are possible.

6
Conclusions

$scCO_2$ has been studied extensively as a versatile medium for carrying out a variety of catalyzed synthetic reactions. The fundamental differences from conventional solvents open up promising avenues for future exploitation. Although $scCO_2$ requires relatively high pressures, non-catalyzed applications in consumer areas, such as natural product extraction (decaffeination), polymer synthesis, and dry cleaning already exist. While there are still limitations to using $scCO_2$ for synthesis, there is no doubt that this medium is beneficial for carrying out these processes and does not have only technological, but also environmental advantages. To exploit the whole potential of $scCO_2$ for catalysis, in particular by using the described switchable properties, the fundamental principles underlying the observed effects must be understood. This needs interdisciplinary research by academic synthetic chemists, physical chemists, chemical engineers, and industrial process chemists. This understanding will later allow for the development of well-defined reaction/separation scenarios that benefit from the unique properties of $scCO_2$.

References

1. Anastas T, Warner J (1998) (eds) Green Chemistry: Theory and Practice. Oxford University Press, Oxford
2. Sheldon RA (2000) Pure Appl Chem 72:1233
3. Anastas PT, Kirchhoff MM (2002) Acc Chem Res 35:686
4. Sheldon RA (1997) Chem Ind, London p 12
5. Jessop PG, Leitner W (1999) Reactions In SCF. In: Jessop PG, Leitner W (eds) Chemical Synthesis Using Supercritical Fluids. Wiley, Weinheim, p 213
6. Clifford AA (1998) Fundamentals of Supercritical Fluids. Oxford University Press, Oxford
7. Zosel K (1978) Angew Chem Int Ed Engl 17:702
8. McHugh MA, Krukonis V (1994) J Supercritical Fluid Extraction: Principles and Practice. Butterworth-Heinemann, Stoneham, MA
9. Rondolph TW (1990) Trends Biotechnol 8:78
10. Gopalan AS, Wai CM, Jacobs HK (2003) (eds) Supercritical Carbon Dioxide: Separations and Processes. Oxford University Press, Washington, DC
11. Eckert CA, Knutson BL, Debenedetti PG (1996) Nature 383:313
12. Cooper AI (2000) J Mater Chem 10:207
13. Field CN, Hamley PA, Webster JM, Gregory DH, Titman JJ, Poliakoff M (2000) J Am Chem Soc 122:2480
14. Esumi K, Sarashina S, Yoshimura T (2004) Langmuir 20:1589
15. Xie B, Finstad CC, Muscat AJ (2005) Chem Mater 17:1753
16. Oakes RS, Clifford AA, Rayner CM (2001) J Chem Soc Perkin Trans 1 917:941
17. Prajapati D, Gohain M (2004) Tetrahedron 60:815
18. Jessop PG, Ikariya T, Noyori R (1995) Science 269:1065
19. Jessop PG, Leitner W (1999) Metal-Complex Catalyzed Reactions. In: Jessop PG, Leitner W (eds) Chemical Synthesis Using Supercritical Fluids. Wiley, Weinheim, p 351
20. Baiker A (1999) Chem Rev 99:453
21. Jessop PG, Ikariya T, Noyori R (1999) Chem Rev 99:475
22. Subramaniam B, Lyon CJ, Arunajatesan V (2002) Appl Catal B 37:279
23. Leitner W (2002) Acc Chem Res 35:746
24. Grunwaldt J-D, Wandeler R, Baiker A (2003) Catal Rev Sci Eng 45:1
25. Supplementary information is available on the homepage of ConNeCat "Competence Network Catalysis" (http://www.connecat.de/)
26. Stanley HE (1987) Phase Transitions and Critical Phenomena. Clarendon Press, Oxford
27. Wesch A, Dahmen N, Ebert KH (1996) Ber Bunsenges Phys Chem 100:1368
28. Raveendraan P, Ikushima Y, Wallen SL (2005) Acc Chem Res 38:478
29. Blanchard LA, Hancu D, Beckman EJ, Brennecke JF (1999) Nature 399:28
30. Aki SNVK, Mellein BR, Saurer EM, Brennecke JF (2004) J Phys Chem B 108:20355
31. Kauffman JF (2001) J Phys Chem A 105:3433
32. Tucker SC (1999) Chem Rev 99:391
33. Kajimoto O (1999) Chem Rev 99:355
34. Fornari RE, Alessi P, Kikic I (1990) Fluid Phase Equilib 57:1
35. Scurto AM (2002) PhD Thesis, University of Notre Dame
36. Dohrn R, Brunner G (1995) Fluid Phase Equilib 106:213
37. Pauchon V, Cissé Z, Chavret M, Jose J (2004) J Supercrit Fluids 32:115

38. Bray CL, Tan B, Wood CD, Cooper AI (2005) J Mater Chem 15(4):456
39. Laugier S, Richon D (1986) Rev Sci Instrum 57:469
40. Fontalba F, Richon D, Renon H (1984) Rev Sci Instrum 55:944
41. Guigard SE, Stiver WH (1998) Ind Eng Chem Res 37:3786
42. Ruckenstein E, Shulgin I (1998) Ind Eng Chem Res 42:1106
43. da Silva MV, Barbosa D (2004) J Supercrit Fluids 31:9
44. Wichterle I (1993) Pure Appl Chem 65:1003
45. Cece A, Jureller SH, Kerschner JL, Moschner KF (1996) J Phys Chem 100:7435
46. Ashour I, Almehaideb R, Fateen S-E, Aly G (2000) Fluid Phase Equilib 167:41
47. Muller EA, Gubbins KE (2001) Ind Eng Chem Res 40:2193
48. Peng D, Robinson DB (1976) Ind Eng Chem Fundam 15:59
49. Zer-Ran Yu, Singh B, Rizvi SH (1994) J Supercrit Fluids 7:51
50. Yang H, Zhong C (2005) J Supercrit Fluids 33:99
51. Chrastil J (1982) J Phys Chem 86:3016
52. Lucien FP, Foster NR (1996) Ind Eng Chem Res 35:4686
53. Ghaziaskar HS, Eskandari H, Daneshfar A (2003) J Chem Eng Data 48:236
54. Mu T, Zhang X, Liu Z, Han B, Li Z, Jiang T, He J, Yang G (2004) Chem Eur J 10:371
55. Yazdi AV, Beckman EJ (1997) Ind Eng Chem Res 36:2368
56. Wagner K, Dahmen N, Dinjus E (2000) Chem Eng Data 45(4):672
57. Raveendran P, Wallen SL (2002) J Am Chem Soc 124:7274
58. Potluri VK, Xu J, Enick R, Beckman E, Hamilton AD (2002) Org Lett 4:14
59. Saffarzadeh-Matin S, Chuck CJ, Kerton FM, Rayner CM (2004) Organometallics 23:5176
60. Yazdi AV, Beckman EJ (1996) Ind Eng Chem Res 35:3644
61. Yee GG, Fulton JL, Smith RD (1992) J Phys Chem 96:6172
62. Dardin A, DeSimone JM, Samulski ET (1998) J Phys Chem B 102:1775
63. Diep P, Jordan KD, Johnson JK, Beckman E (1998) J Phys Chem A 102:2231
64. Yonker CR, Palmer BJ (2001) J Phys Chem A 105:308
65. Kanakubo M, Umecky T, Liew CC, Aizawa T, Hatakeda K, Ikushima Y (2002) Fluid Phase Equilib 194:859
66. Raveendran P, Wallen SL (2003) J Phys Chem B 107:1473
67. Laintz KE, Wai CM (1991) J Supercrit Fluids 4:194
68. Esumi K, Sarashina S, Yoshimura T (2004) Langumir 20:1589
69. Wang JS, Wai CM (2005) Ind Eng Chem Res 44:922
70. Giles MR, Hay JN, Howdle SM, Winder RJ (2000) Polymer 41:6715
71. Khosravi-Darani K, Vasheghani-Farahani E, Yamini, Y, Bahramifar N (2003) J Chem Eng Data 48:860
72. Giles MR, Hay JN, Howdle SM (2000) Macromol Rapid Commun 21:1019
73. Kani I, Omary MA, Raeashedeh-Omary MA, Lopez-Castillo ZK, Flores R, Akgerman A, Fackler JP Jr (2002) Tetrahedron 58:3923
74. Wright ME, Lott KM, McHugh MA, Shen Z (2003) Macromolecules 36:2242
75. Saraf MK, Gerard S, Wojinski LM, Charpentier PA, Desimone JM, Roberts GW (2002) Macromolecules 35:7976
76. Sarbu T, Styranec T, Beckman EJ (2000) Nature 405:165
77. Laintz KE, Wai CM, Yonker CR, Smith RD (1992) Anal Chem 64:2875
78. Smith DC, Stevens ED Jr, Nolan SP (1999) Inorg Chem 38:5277
79. Chen W, Xiao J (2000) Org Lett 2:2676
80. Chen W, Hu Y, Osuna AMB, Xiao J (2002) Tetrahedron 58:3889
81. Hu Y, Chen W, Xu L, Xiao J (2001) Organometallics 20:3206

82. Montilla F, Rosa V, Prevett C, Avilés T, Nunes da Ponte M, Masi D, Mealli C (2003) Dalton Trans, p 2170
83. Montilla F, Galindo A, Rosa V, Avilés T (2004) Dalton Trans, p 2588
84. Wai CM, Wang S (2000) J Biochem Methods 43:273
85. Cowey CM, Bartle KD, Burford MD, Clifford AA, Zhu S, Smart NG, Tinker ND (1995) J Chem Eng Data 40(6):1217
86. Kreher U, Schebesta S, Walther D (1998) Z Anorg Allg Chem 624:602
87. Burk MJ, Feng S, Gross MF, Tumas W (1995) J Am Chem Soc 117:8277
88. Jessop PG, Ikariya T, Noyori R (1995) Organometallics 14:1510
89. Guzel B, Omary MA, Fackler JP, Akgerman A (2001) Inorg Chim Act 325:45
90. Zhao F, Ikushima Y, Chatterjee M, Sato O, Arai M (2003) J Supercrit Fluids 27:65
91. Hua Y, Birdsall DJ, Stuart AM, Hope EG, Xiao J (2004) J Mol Catal A:Chem 219:57
92. Dong X, Erkey C (2004) J Mol Catal A:Chem 211:73
93. Lange S, Brinkmann A, Trautner P, Woelk K, Bargon J, Leitner W (2000) Chirality 12:450
94. Carroll MA, Holmes AB (1998) Chem Commun, p 1395
95. Early TR, Gordon RS, Carroll MA, Holmes AB, Shute RE, McConvey IF (2001) Chem Commun, p 1966
96. Osswald T, Schneider S, Wang S, Bannwarth A (2001) Tetrahedron Lett 42:2965
97. Shezad N, Clifford AA, Rayner CM (2002) Green Chem 4:64
98. Bhanage BM, Fujita S, Arai M (2003) J Organomet Chem 687:211
99. Hamilton JG, Rooney JJ, DeSimone JM, Mistele C (1998) Macromolecules 31:4387
100. Fürstner A, Ackermann L, Beck K, Hori H, Koch D, Langemann K, Liebl M, Six C, Leitner W (2001) J Am Chem Soc 123:9000
101. Song RQ, Zeng JQ, Zhong B (2002) Catal Lett 82:89
102. Jeong N, Hwang SH, Lee YW, Lim JS (1997) J Am Chem Soc 119:10549
103. Wegner A, Leitner W (1999) Chem Commun, p 1583
104. Montilla F, Avile's T, Casimiro T, Ricardo AA, da Ponte MN (2001) J Organomet Chem 632:113
105. Cheng J-S, Jiang H-F (2004) Eur J Org Chem, p 643
106. de Vries TJ, Duchateau R, Vorstmana MAG, Keurentjes JTF (2000) Chem Commun, p 263
107. Hori H, Six C, Leitner W (1999) Macromolecules 32:3178
108. Carter CAG, Baker RT, Nolan SP, Tumas W (2000) Chem Commun, p 347
109. Cacchi S, Fabrizi G, Gasparrini F, Pace P, Villani C (2000) Synlett, p 650
110. Wittmann K, Wisniewski W, Mynott R, Leitner W, Kranemann CL, Rische T, Eilbracht P, Kluwer S, Ernsting JM, Elsevier CJ (2001) Chem Eur J 7:4584
111. Birnbaum ER, Le Lacheur RM, Horton AC, Tumas W (1999) J Mol Catal A:Chem Chem 139:11
112. Kokubo Y, Wu X-W, Oshima Y, Koda S (2004) J Supercrit Fluids 30:225
113. Jessop PG, Ikariya T, Noyori R (1994) Nature 368:231
114. Jessop PG, Hsiao Y, Ikariya T, Noyori R (1996) J Am Chem Soc 118:344
115. Rohr M, Geyer C, Wandeler R, Schneider M-S, Murphy EF, Baiker A (2001) Green Chem 3:123
116. Schmid L, Schneider MS, Engel D, Baiker A (2003) Catal Lett 88:105
117. Dahmen N, Griesheimer P, Makarczyk P, Pitter S, Walter O (2005) J Organom Chem 690(6):1467
118. Bhattacharyya P, Gudmunsen D, Hope EG, Kemmit RDW, Paige DR, Stuart AM (1997) J Chem Soc: Perkin Trans 1, p 3609

119. Horváth IT, Rabai J (1994) Science 266:72
120. Richter B, de Wolf E, van Koten G, Deelman BJ (2000) J Org Chem 65:3885
121. Gladysz JA, Corrêa da Costa R (2004) Strategies for the Recovery of Fluorous Catalysts and Reagents: Design and Evaluation. In: Gladysz JA, Curran DP, Horváth IT (eds) Handbook of Fluorous Chemistry. Wiley, Weinheim, p 24
122. Palo D, Erkey C (1998) J Chem Eng Data 43:47
123. Chen MJ, Klingler RJ, Rathke JW, Kramarz KW (2004) Organometallics 23:2701
124. Aschenbrenner O, Ionescu C, Makarczyk P, Pitter S (2005) (in preparation)
125. Ionesu C, Makarczyk P, Pitter S (2005) (in preparation)
126. Leadbeater NE, Marco M (2002) Chem Rev 102:3217
127. Lu Z, Lindner E, Mayer HA (2002) Chem Rev 102:3543
128. van Heerbeek R, Kamer PCJ, van Leeuwen PWNM, Reek JNH (2002) Chem Rev 102:3717
129. Cornils B, Herrmann WA (2004) (eds) Aqueous-Phase Organometallic Catalysis, 2nd edn. Wiley, Weinheim
130. Welton T (1999) Chem Rev 99:2071
131. Brasse CC, Englert U, Salzer A, Waffenschmidt H, Wasserscheid P (2000) Organometallics 19:3818
132. Dupont J, de Souza RF, Suarez PAZ (2002) Chem Rev 102:3543
133. Horvath IT (1998) Acc Chem Res 31:641
134. Betzemeier B, Knochel P (1999) Top Curr Chem 206:61
135. Kainz S, Koch D, Baumann W, Leitner W (1997) Angew Chem Int Ed Engl 36:1628
136. Koch D, Leitner W (1998) J Am Chem Soc 120:13398
137. Kainz S, Brinkmann A, Leitner W, Pfaltz A (1999) J Am Chem Soc 121:6421
138. Baker RT, Tumas W (1999) Science 284:1477
139. Bergbreiter DE (2002) Chem Rev 102:3345
140. Cole-Hamilton DJ (2003) Science 299:1702
141. Keim W (2003) Green Chem 5:105
142. Frohning CD, Kohlpaintner CW, Bohnen H-W (2002) Hydroformylation (Oxo Synthesis, Roelen Reaction) In: Cornils B, Herrmann WA (eds) Applied homogeneous catalysis with organometallic compounds, vol 1. Wiley, Weinheim, p 31
143. Hâncu D, Beckman E (2001) Green Chem 3:80
144. Hyde JR, Licence P, Carter D, Poliakoff M (2001) Appl Catal A:Chem 222:119
145. Webb PB, Sellin MF, Kunene TE, Williamson S, Slawin AMZ, Cole-Hamilton DJ (2003) J Am Chem Soc 125:15577
146. Goetheer ELV, Verkerk AW, van den Broeke LJP, de Wolf E, Deelman B-J, van Koten G, Keurentjes JTF (2003) J Catal 219:126
147. Webb PB, Cole-Hamilton DJ (2004) Chem Commun, p 612
148. Hyde JR, Poliakoff M (2004) Chem Commun, p 1482
149. Walsh B, Hyde JR, Licence P, Poliakoff M (2005) Green Chem 7:456
150. Kokubo Y, Hasegawa A, Kuwata S, Ishihara K, Yamamoto H, Ikariyaa T (2005) Adv Synth Catal 347:220
151. Roelen O (1938) Ger Patent 949 548
152. Beller M, Cornils B, Frohning DC, Kohlpaintner CV (1995) J Mol Catal A: Chem 104:17
153. Bohnen H-W, Cornils B (2002) Adv Catal 47:1
154. Heck RF, Breslow DA (1961) J Am Chem Soc 83:4023
155. Rathke JW, Klingler RJ, Krause TR (1991) Organometallics 10:1350
156. Rathke JW, Klingler RJ, Krause TR (1992) Organometallics 11:585

157. Cross W Jr, Akgerman A, Erkey D (1996) Ind Eng Chem Res 35:1765
158. Guo Y, Akgerman A (1997) Ind Eng Chem Res 36:4581
159. Kainz S, Koch D, Baumann W, Leitner W (1997) Angew Chem Int Ed Engl 36:1628
160. Bach I, Cole-Hamilton DJ (1998) Chem Commun, p 1463
161. Palo DR, Erkey C (1998) Ind Eng Chem Res 37:4203
162. Guo Y, Akgerman A (1999) J Supercrit Fluid 15:63
163. Palo DR, Erkey C (1999) Ind Eng Chem Res 38:2163
164. van Leeuwen PWNM, Sandee AJ, Reek JNH, Kamer PCJ (2002) J Mol Catal A 182–183:107
165. Fujita S, Fujisawa S, Bhanagea BM, Arai M (2004) Tetrahedron Lett 45:1307
166. Jessop PG, Wynne DC, DeHaai S, Nakawatase D (2000) Chem Commun 693
167. Palo DR, Erkey C (2000) Organometallics 19:81
168. Sellin MF, Cole-Hamilton DJ (2000) J Chem Soc Dalton Trans, p 1681
169. Meehan NJ, Sandee AJ, Reek JNH, Kamer PCJ, van Leeuwen PWNM, Poliakoff M (2000) Chem Commun, p 1497
170. Dharmidhikari D, Abraham MA (2000) J Supercrit Fluids 18:1
171. Sandee AJ, Reek JNH, Kamer PCJ, van Leeuwen PWNM (2001) J Am Chem Soc 123:8468
172. Snyder G, Tadd A, Abraham MA (2001) Ind Eng Chem Res 40:5317
173. Tadd AR, Marteel A, Mason MR, Davies JA, Abraham MA (2003) J Supercrit Fluids 25(2):183
174. Patcas F, Maniut C, Ionescu C, Pitter S, Dinjus E (2005) Appl Catal B, (in press)
175. Chaudhari RV, Bhanage BM, Deshpande RM, Delmas H (1995) Nature 373:501
176. Wiese K-D, Möller O, Protzmann P, Trocha M (2003) Catal Today 79–80:97
177. Chaudhari RV, Mills PL (2004) Chem Eng Sci 59:5337
178. Pruett RL, Smith JA (1968) Patent ZA 6,804,937; NL 68-10783
179. Cornils B, Hermann WA (eds) (2002) Applied Homogeneous Catalysis with Organometallic Compounds, 2nd edn. Wiley, Weinheim
180. Cents AHG, Brilman DHF, Versteeg GF (2004) Ind Eng Chem Res 43:7465
181. Kramer GM, Leder F (1975) US Patent 3,880,945
182. Pereda S, Bottini SB, Brignole EA (2005) Appl Catal A:Chem 281:129
183. Subramaniam B, McHugh MA (1986) Ind Eng Chem Process Des Dev 25:1
184. Tiltscher H, Hofmann H (1987) Chem Eng Sci 42:959
185. Dahl S, Michelsen ML (1990) AIChE J 36:1829
186. Klinger RJ, Rathke JW (1994) J Am Chem Soc 116:4772
187. Savage P, Gopalan S, Mizan T, Martino C, Brock E (1995) AIChE J 41:1723
188. Bertuco A, Canu P, Devetta L, Zwahlen AG (1997) Ind Eng Chem Res 36:2626
189. Kramarz KW, Klingler RJ, Fremgen DE, Rathke JW (1999) Catal Today 49(4):339
190. Fujita S, Fujisawa S, Bhanage BM, Ikushima Y, Arai M (2002) New J Chem 26:1479
191. Pino P, Piacenti F, Bianchi M, Lazzaroni R (1968) Chim Ind (Milan) 50:106
192. Parshall GW, Ittel SD (1992) Homogeneous Catalysis, 2nd edn. Wiley, New York
193. Hu Y, Chen W, Banet Osuna AM, Stuart AM, Hope EG, Xiao J (2001) Chem Commun, p 725
194. Hu Y, Birdsall DJ, Stuart AM, Hope EG, Xiao J (2004) J Mol Catal A:Chem 219:57
195. van Eldik R, Hubbard CD (eds) (1997) Chemistry Under Extreme or Non-Classical Conditions. Wiley and Spektrum Akademischer Verlag co-publication, New York
196. Savage P (1999) Chem Rev 99:603
197. Darr JA, Poliakoff M (1999) Chem Rev 99:495

198. Jessop PG, Leitner W (eds) (1999) Chemical Synthesis Using Supercritical Fluids. Wiley, Weinheim, p 388
199. Sandee AJ, van der Veen LA, Reek JNH, Kamer PCJ, Lutz M, Spek AL, van Leeuwen PWNM (1999) Angew Chem Int Ed Engl 38:3231
200. Blanchard LA, Hancu D, Beckman EJ, Brennecke JF (1999) Nature 399:28
201. Blanchard LA, Gu ZY, Brennecke JF (2001) J Phys Chem B 105:2437
202. Anthony JL, Maginn EJ, Brennecke JF (2002) J Phys Chem B 106:7315
203. Cadena C, Anthony JL, Shah JK, Morrow TI, Brennecke JF, Maginn EJ (2004) J Am Chem Soc 126:5300
204. Cole-Hamilton D (2005) Green Chem 7:373
205. Blanchard LA, Brennecke JF (2001) Ind Eng Chem Res 40:287
206. Scurto AM, Aki S, Brennecke JF (2002) J Am Chem Soc 124:10276
207. Sovova H, Jez J (1994) J Chem Eng Data 39(4):840
208. Francio G, Wittmann K, Leitner W (2001) J Organomet Chem 621:130
209. Bahrmann H, Bach H (2002) Ullmann's Encyclopedia of Industrial Chemistry, "Oxo-Synthesis". CEElectronic Release, 6th ed., Wiley, Weinheim, p 1
210. Dwyer C, Assumption H, Coetzee J, Crause C, Damoense L, Kirk M (2004) Coord Chem Rev 248:653
211. Manning AR (1968) J Chem Soc 1135
212. Farrar DH, Poe J, Stromnova TA (1997) Organometallics 16(13):2827
213. Licence P, Ke J, Sokolova M, Ross SK, Poliakoff M (2003) Green Chem 5:99
214. Tominaga K, Sasaki Y (2000) Catal Commun 1:1
215. Tominaga K, Sasaki Y (2004) J Mol Catal A:Chem 220(2):159

Top Organomet Chem (2008) 23: 149–161
DOI 10.1007/3418_042
© Springer-Verlag Berlin Heidelberg
Published online: 15 July 2006

Catalytic SILP Materials

Anders Riisager[1] (✉) · Rasmus Fehrmann[1] · Marco Haumann[2] ·
Peter Wasserscheid[2] (✉)

[1]Department of Chemistry and Center for Sustainable and Green Chemistry,
Technical University of Denmark, Building 207, 2800 Kgs. Lyngby, Denmark
ar@kemi.dtu.dk

[2]Lehrstuhl für Chemische Reaktionstechnik (CRT), Universität Erlangen-Nürnberg,
Egerlandstrasse 3, 91058 Erlangen, Germany
wasserscheid@crt.cbi.uni-erlangen.de

Abstract The principle of catalytic SILP materials involves surface modification of a porous solid material by an ionic liquid coating. Ionic liquids are salts with melting points below 100 °C, generally characterized by extremely low volatilities. In the examples described in this paper, the ionic liquid coating contains a homogeneously dissolved Rh-complex and constitutes a uniform, thin film, which itself displays the catalytic reactivity in the system. Continuous fixed-bed reactor technology has been applied successfully to demonstrate the feasibility of catalytic SILP materials for propene hydroformylation and methanol carbonylation.

Keywords Carbonylation · Homogeneous catalysis · Hydroformylation · Immobilisation · Ionic liquids · Supported catalysts

1
Introduction

Natural and synthesized solid materials are generally characterized by a non-uniform and undefined surface. The surface contains face atoms, corner atoms, edge atoms, ad-atoms and defect sites, which all together determine the surface properties of the material. In many applications, these different sites display different properties, e.g., with respect to their chemical activity. Often, only certain sites are advantageous with regard to e.g., the reactivity or selectivity of the materials if used as a heterogeneous catalyst. Consequently, the design of solid materials with greatly enhanced surface uniformity and specificity is a task of great relevance for the whole field of material science and especially for heterogeneous catalysis.

The principle of the supported ionic liquid phase (SILP) technology involves surface modification of a solid material by an ionic liquid coating. Ionic liquids are salts with melting points below 100 °C [1–8] generally characterized by extremely low volatilities under ambient conditions [9]. The ionic liquid coating constitutes a thin film, which is confined on the surface of the solid by various methods such as, e.g., physisorption, tethering, or covalent anchoring of ionic liquid fragments [10]. Ionic liquids are further characterized by containing a highly pre-organized, homogeneous liquid structure with distinctive physicochemical characteristics. These—often unique—characteristics are exclusively governed by the combination of the specific ions in the material. Hence, by an appropriate choice of the ions contained in the ionic liquid material, it is possible to transfer specific properties of the fluid to the surface of a solid material. In the case of catalytic SILP materials, the SILP concept allows tailor-made catalysts by confining a catalytically active, ionic liquid-transition metal complex solution onto the surface. Figure 1 schematically illustrates such a catalytic SILP material containing Rh-complexes dissolved in the ionic liquid film.

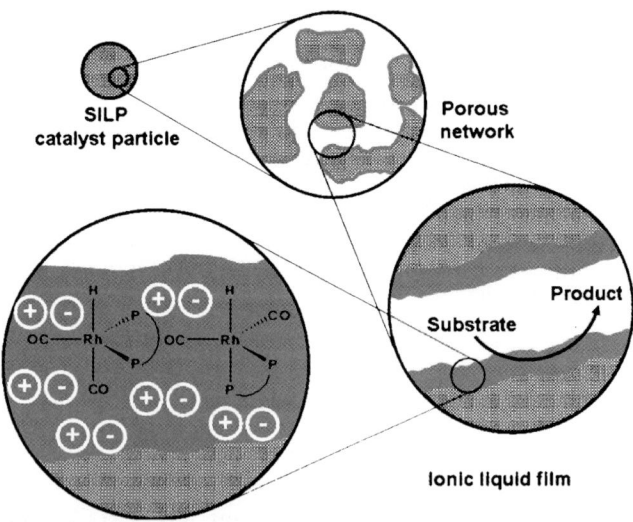

Fig. 1 Schematic presentation of supported ionic Liquid Phase (SILP) catalyst

Apart from the fact that the ionic liquid covers the more or less undefined surface of the support to enhance uniformity of the material, the concept makes use of the fact that ionic liquids have been found in many studies to be highly suitable solvents for transition metal catalyzed reactions [11–15]. The almost unlimited combinations of cation and anion further allow the synthesis of tailor made ionic liquids that can stabilize the catalytic species [16–18]. Traditionally, ionic liquids have been used in transition metal-catalyzed reactions using liquid-liquid systems comprising an ionic liquid and an immiscible organic solvent. Even if excellent catalyst immobilization and thus high degrees of recyclability of the ionic liquid catalyst solution have been demonstrated for many applications, the commercialization of this approach is clearly hampered by the large amounts of ionic liquid required [19]. This is due to the fact that ionic liquids remain expensive solvents, even though being widely commercially available by now [20–26]. In addition, the high viscosity of ionic liquids can induce mass transfer limitations in these liquid–liquid biphasic systems if the chemical reaction is fast, in which case the reaction takes place only within a narrow diffusion layer and not in the main part of the ionic liquid catalyst solution. Thus only a minor part of the ionic liquid and dissolved precious transition metal catalyst are utilized efficiently.

In a number of publications we have recently demonstrated that this problem of mass transport limitation can be circumvented by using catalytic SILP materials [27–32]. Moreover, these catalysts allow the application of fixed-bed reactors for simple continuous processing when applied in combination with gaseous reaction mixtures making the separation and catalyst recycling obsolete.

The use of catalytic SILP materials has been reviewed recently [10] covering Friedel-Crafts reactions [33–37], hydroformylations (Rh-catalyzed) [38], hydrogenation (Rh-catalyzed) [39, 40], Heck reactions (Pd-catalyzed) [41], and hydroaminations (Rh-, Pd-, and Zn-catalyzed) [42]. Since then, the SILP concept has been extended to additional catalytic reactions and alternative support materials. In this paper we will present results from continuous, fixed-bed carbonylation and hydroformylation reactions using rhodium-based SILP catalysts as reaction examples demonstrating the advantages of the SILP technology for bulk chemical production.

2
Experimental

All syntheses were carried out using standard *Schlenk* techniques under argon. Support materials were purchased from Merck (SiO$_2$, Al$_2$O$_3$), Millennium Co. (TiO$_2$) and MEL Chemicals (ZrO$_2$). Rh(CO)$_2$(acac) (98%) was purchased from Aldrich, while [Rh(CO)$_2$I]$_2$ and phosphine ligands were prepared according to literature procedures. All of these materials were used without further purification. Ionic liquids were purchased from Solvent Innovation [24] and dried at high vacuum overnight prior to use.

SILP hydroformylation catalysts were prepared by dissolving appropriate amounts of $Rh(CO)_2(acac)$ and ligand in dried methanol followed by stirring for 30 min. Afterwards, the ionic liquid was added to the solution. After stirring for another 30 min, the support was added and the solution stirred for 60 min. Finally, the methanol was removed in vacuo and a pale red powder was obtained. The supported ionic liquid-phase catalyst was stored under argon for further use.

The SILP carbonylation catalyst was prepared by one-step impregnation of silica support using a methanolic solution of the ionic liquid [BMIM]I and the dimer $[Rh(CO)_2I]_2$. The use of the dimeric precursor complex allowed formation of the catalyst anion $[Rh(CO)_2I_2]^-$ directly during catalyst preparation without formation of contaminating byproducts in the ionic liquid catalyst solution.

3
Catalytic SILP Materials in Hydroformylation

In a set of preliminary experiments the influence of each single SILP component has been studied. The support type was found to have the most profound effect on SILP catalyst performance. The phosphine ligand excess was reduced due to Brønsted acidic surface sites of SiO_2. A thermal pre-treatment of the silica supports, by which the surface silanol groups were removed, was found to be a prerequisite for good catalyst performance. The silica supports had residual surface acidity even after thermal pre-treatment, thus the L/Rh ratio had to be adjusted to 10. When using less acidic supports like TiO_2, already the untreated material gave good results. The influence of different phosphine ligands on hydroformylation activity and selectivity was tested. These ligands are based on triphenylphosphine 1, bridged phosphine 2 (both monodentate) and diphenyl-xantphos 3 (bidentate) backbones, as depicted in Fig. 2.

For the monophosphine catalysts the obtained selectivities proved significantly lower than for analogous bisphosphine catalysts as expected. Moreover, an increase in monophosphine ligand content generally resulted in less active but more selective SILP catalysts. Accordingly, the optimal L/Rh ratio was found to be around 10 for both the monophosphine and the bisphospine modified catalysts. The ionic liquid loading α (volume of ionic liquid per pore volume of support) had also a direct influence on the catalyst activity. At higher liquid loadings the activity decreased due to pore filling of the support, which resulted in decreased interfacial reaction area. The effect of the ionic liquid anion on catalyst performance was found to be insignificant. Similar results were obtained when using [BMIM][PF_6]

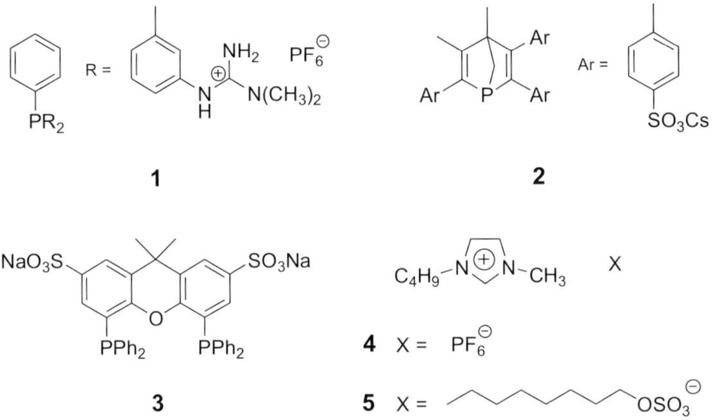

Fig. 2 Phosphine ligands and ionic liquids used in continuous hydroformylation of propene [32]

(4) and [BMIM][n-C$_8$H$_{17}$O – SO$_3$] (5). In consecutive experiments, catalysts based on 5 were applied, and the ionic liquid loading was maintained at $\alpha = 0.1$ in order to ensure high liquid dispersion and thereby large reaction surface area.

3.1
Long-Term Stability Studies

All the preliminary measurements of the Rh-3/5/SiO$_2$ (denoted as Rh-3-SILP) catalyst discussed so far were initially studied at 100 °C and 10 bar syngas pressure (H$_2$: CO 1 : 1) over a period of up to 36 h. This time on stream was further extended to 180 hours to test the long-term stability of the Rh-3-SILP dehydroxylated catalyst system (Fig. 3).

In the extended reaction period, the selectivity towards linear butanal remained constant around 95% ($n/iso = 19$), whereas the TOFs slightly decreased over time, corresponding to a total loss in activity of 17% or 0.1% per hour. Since the selectivity remained unchanged it was excluded that deactivation was caused by catalyst decomposition, as observed in the previous studies when using low L/Rh ratios. ICP analyses of the exit gas streams condensed by liquid nitrogen further established the rhodium contents to be below the detection limit of 3 ppm. Instead, the formation of high boiling side-products, dissolving in the ionic liquid layer during reaction causing a lowering in the effective rhodium concentration, was suspected to be the reason of the slow activity decrease measured. Furthermore, the film thickness might be increased and smaller pores flooded with the formation of high-boiling side products, which would lead also to a lower reaction surface. To confirm this hypothesis, the gas flow was temporary

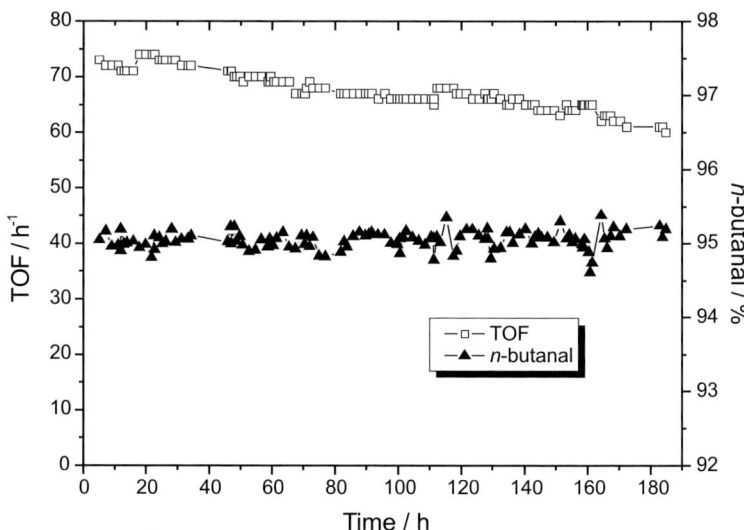

Fig. 3 Long-term stability of Rh-**3**-SILP catalyst in continuous gas-phase hydroformylation of propene. (Reaction conditions: $T = 100\,^\circ$C, $p_{total} = 10$ bar with $p_{H_2} = p_{CO} = 4.1$ bar and $p_{propene} = 1.8$ bar, $n_{rhodium} = 3.53 \times 10^{-5}$ mol, $\tau = 0.38$ s) [31]

stopped after 180 h time on stream and the reaction setup was evacuated for 10 min at $100\,^\circ$C. When the experiment was continued after this procedure, the activity had indeed increased by 80% from TOFs of initially $60\,h^{-1}$ to $108\,h^{-1}$. Within the next 20 h of reaction the TOFs decreased again from $108\,h^{-1}$ to $76\,h^{-1}$ and the selectivity re-established at 95% *n*-butanal. A second vacuum period of 10 min resulted in improved TOFs, as depicted in Fig. 4. In both cases the observed "overshooting" of the activity directly after evacuation might be caused by either simultaneous removal of CO ligand of the Rh-**3**-complex leading to higher activity or a rearrangement of the active surface due to sudden evaporation of dissolved heavies. In the first case, a lower selectivity would be expected, which was indeed observed directly after the evacuation. Thereafter, the catalyst solution was re-saturated with CO gas and both the activity and the selectivity approached the initial levels.

The findings from the evacuation experiment confirmed the interpretation that the observed slight deactivation over time was not due to catalyst decomposition, as an accompanying decrease in activity and selectivity otherwise would have been observed. Instead, we expected the formation of aldol products 2-ethyl-hexanal and 2-ethyl-hexanol to be of relevance, as traces of these high boiling side products were observed at particular high conversions.

To support this interpretation the deactivated pale yellow catalyst was extracted by cyclohexane and ethanol, respectively. Immediately, the ethanol solution became orange whereas the cyclohexane solution remained color-

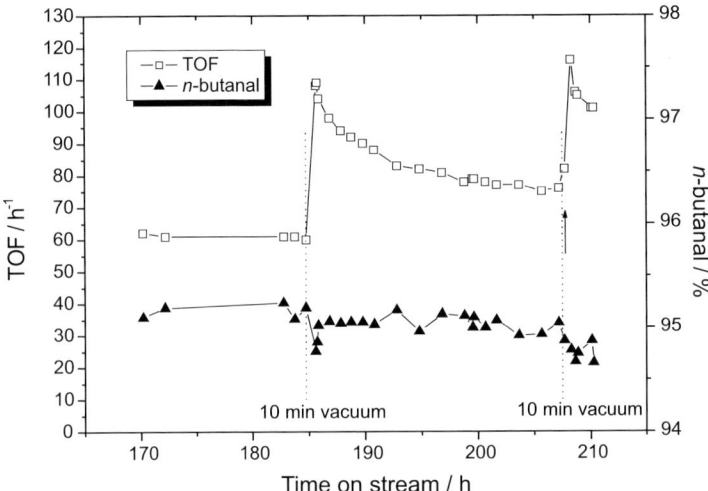

Fig. 4 Reactivation of Rh-3-SILP catalyst by consecutive application of vacuum (Reaction conditions: $T = 100\,°C$, $p_{total} = 10$ bar with $p_{H_2} = p_{CO} = 4.1$ bar and $p_{propene} = 1.8$ bar, $n_{rhodium} = 3.53 \times 10^{-5}$ mol, $\tau = 0.38$ s) [32]

less, indicating that only the ethanol wash led to removal of the ionic catalyst phase from the support. Both solutions were subsequently analyzed by means of GC-MS (ethanol solution after distillation and separation of non-volatile ionic liquid fragments), where content of small amounts of by-products 1-butanol, 2-ethyl-hexanal and 2-ethyl-hexanol were confirmed, as expected. Additionally, infrared analysis of both used and unused Rh-3-SILP catalysts was carried out. The used catalyst showed an absorbance band at $1730\,cm^{-1}$ (corresponding to the $C = O$ stretching frequency) clearly indicating the presence of aldehydes in the catalyst.

3.2
Kinetic Studies

To verify the homogeneous nature of Rh-3-SILP catalysts, as previously suggested based on IR and NMR spectroscopic studies, [30] kinetic experiments have also been conducted with the catalyst. Here, a continuous fixed-bed reactor setup equipped with online gas-chromatography, described elsewhere in detail, [31] was applied. The general rate law for the hydroformylation of propene was assumed:

$$r = k \cdot p_{propene}^{n} \cdot p_{H_2}^{x} \cdot p_{CO}^{y} \tag{1}$$

In the first kinetic experiments, the substrate concentration was varied by changing the propene partial pressure in the feed gas (at constant total pressure) between 0.9–3.2 bar at temperatures in the range 65–140 °C. The rate constants k were derived from differential analyses of the data, and the reaction order with respect to propene partial pressure was determined to be first order, i.e., $n = 1$ in Eq. 1. This dependency was similar to traditional organic and aqueous Rh-phosphine catalyzed hydroformylation systems [43].

Next, the residence time inside the catalyst bed was varied by altering the total reactant flow at a constant propene partial pressure of 2.1 bar. The selectivity for the desired linear aldehyde was only influenced by temperature and not by residence time. Shorter residence times generally resulted in lower conversions, as expected. However, under non-differential conditions at higher conversions the observed TOFs decreased slightly with longer residence times, due to lower mean levels of propene present in the reactor.

From the independent propene pressure and residence time experiments the activation energy was calculated from Arrhenius plots to be $63.3 \pm 2.1 \text{ kJ mol}^{-1}$. This value clearly indicated that the Rh-3-SILP catalyst was operating under kinetically controlled reaction conditions and established the supposition of the homogeneous nature of the Rh-3-SILP catalyst as confirmed by spectroscopic studies (vide supra).

3.3
Variation in Syngas Composition and Total Pressure

In order to further assess the effect of parametrical changes to the Rh-3-SILP hydroformylation system, the ratio between the partial pressures of hydrogen and carbon monoxide ($p_{H_2} : p_{CO}$ ratio) has been varied between 0.25 and 4 (at constant total pressure) in reactions performed at 65 and 100 °C [31]. Increasing the hydrogen partial pressure had a profound effect on the catalyst activity for both temperatures, as depicted in Fig. 5.

The positive effect of a high relative hydrogen pressure is well described by the generally accepted Wilkinson's mechanism for homogeneous ligand modified rhodium hydroformylation [44–46] in which the pre-equilibrium between the dimer and the monomer of the Rh-3-complex is shifted towards the active monomer at higher partial pressures of hydrogen. Furthermore, the observed negative influence of high carbon monoxide partial pressures was also in accordance with this mechanism, in which the formation of catalytically active species requires the loss of CO during the cycle [47, 48]. Thus, these findings provided further verification of the homogeneous nature of the Rh-3-SILP catalyst system.

At 100 °C the catalyst selectivity for n-butanal was practically independent of the $H_2 : CO$ ratio ($< 0.5\%$ variation), while the syngas ratio significantly influenced the selectivity at the lower temperature of 65 °C providing high selectivity of about 98% with a $H_2 : CO$ ratio of 4. Moreover, at 65 °C with a large excess of CO gas ($H_2 : CO = 0.25$) n-butanal was formed exclusively with 99.5% selectivity ($n/iso = 193$). Such high selectivity has not previously

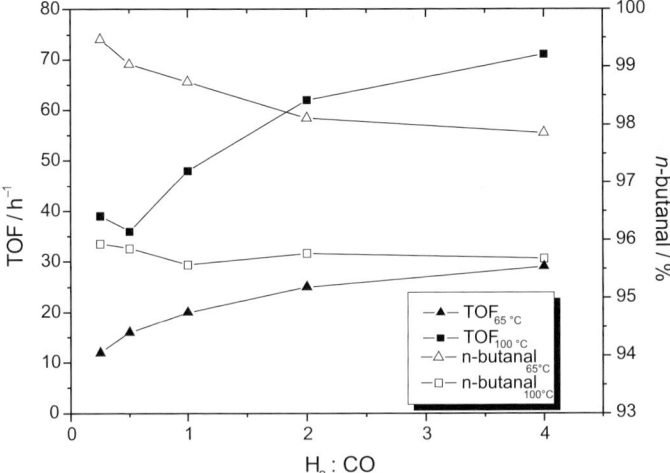

Fig. 5 Activity of Rh-3-SILP catalyst at different syngas compositions (p_{H_2} : p_{CO} ratios) and temperatures (Reaction conditions: $T = 100\,°C$, $p_{total} = 10\,bar$) [32]

been reported for Rh-3 catalyst systems in either homogeneous or biphasic reaction systems, normally performed in batch autoclaves. Hence, these results clearly signify the high potential of the SILP concept for obtaining a better mechanistic understanding in organometallic catalysis, as exceptionally low steady state conversions—impossible to realize in batch mode—can be adjusted in fixed-bed reactors. Additionally, the fixed-bed SILP catalytic concept allows a variety of reaction parameters to be altered over a broad range by using only small amounts of precious catalyst.

4
Catalytic SILP Materials in Carbonylation

In order to demonstrate a more general applicability of the SILP catalyst concept to organometallic catalytic processes, a SILP SiO_2 – [BMIM] – $[Rh(CO)_2I_2]$ – [BMIM]I catalyst system has been developed and applied for continuous, fixed-bed gas-phase methanol carbonylation at industrially relevant reaction conditions [49]. The SILP rhodium-iodide catalyst resemble the widely applied homogeneous Monsanto catalyst system containing the complex anion $[Rh(CO)_2I_2]^-$. Importantly, reactions with the SILP Monsanto catalyst demand less catalyst material and allows a simple process design avoiding recirculation and pressure change of the catalyst system, compared to present carbonylation technologies.

The SILP catalyst system was applied for water-free gas-phase carbonylation of methanol at two different gas space velocities, using a continuous fixed-bed reactor setup equipped with online gas-chromatography. In Fig. 6 the catalyst activity (as TOFs) for formation of the desired acetyl products, acetic acid and methyl acetate, and byproduct dimethyl ether are shown for the first 1.5 h of reaction with low gas velocity at 20 bar pressure.

Here, essentially complete conversion of methanol was obtained with a TOF for acetyl products of 77 h^{-1} and space time yield (i.e., production rate) of 21 mol L^{-1} h^{-1}, however, with a high selectivity towards the ester relative to the acid (ester/acid ratio = 3.5) and ether byproduct. Noteable, the production rate was practically the same as previously obtained using an bubble-column reaction system (water-free) containing about 100 times the volume of ionic liquid catalyst solution, clearly signifying the efficiency of the highly dispersed ionic liquid catalyst layer in the SILP catalyst. The selectivity towards acetic acid (about 21%) was considerably lower than what previously was found for the large-volume reaction system (up to 96% acetic acid), most likely as a result of a relatively longer residence time of the reactant gas in the SILP system. Accordingly, reaction at low space velocity at a pressure of 10 bar (Table 3, entry 3) resulted in increased formation of acetic acid (and ether, 9%) relatively to ester (ester/acid ratio of 5.4) than found at twice the gas space velocity (entry 2, ester/acid ratio = 19.8). Moreover, the byproduct formation was reduced to about 2% by applying a combination of low reaction pressure and long contact time of the gas with the SILP catalyst phase (i.e., low gas space velocity), thus providing a mixed acetyl reaction product with high purity (about 98%), which can be further hydrolyzed into acetic acid if required.

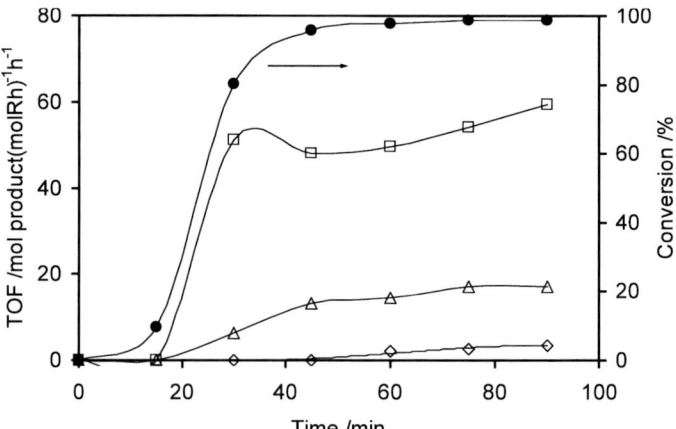

Fig. 6 Methanol conversion and TOFs for product formation using SILP SiO$_2$ – [BMIM][Rh(CO)$_2$I$_2$] – [BMIM]I catalyst in continuous, gas-phase methanol carbonylation (Reaction conditions: $T = 180\,°C$, $P = 20\,bar$, $F_{CO} = 50\,ncm^3\,min^{-1}$, F_{liquid} (MeOH:MeI = 75:25 wt%) = 0.69 g h^{-1}) [49]

5
Conclusion

We have demonstrated the long-term stability of a homogeneous rhodium hydroformylation catalyst, which has been immobilized by the use of the new supported ionic liquid phase (SILP) technique. No loss in selectivity for linear butanal was observed during the experiments. The activity decreased by 17% within 180 h time-on-stream due to formation of high boiling side-products. These heavies could easily be removed from the catalyst by a vacuum procedure after which the initial activity could be regained. The activation energy of 63.3 ± 2.1 kJ mol^{-1} added evidence that the catalyst is indeed a homogeneous complex dissolved in an ionic liquid film on a support. Furthermore, the Rh-SILP catalyst performed very similar to a homogeneous catalyst with regard to variation in syngas composition.

In SILP carbonylation we have introduced a new methanol carbonylation SILP Monsanto catalyst, which is different from present catalytic alcohol carbonylation technologies, by using an ionic liquid as reaction medium and by offering an efficient use of the dispersed ionic liquid-based rhodium-iodide complex catalyst phase. In perspective the introduced fixed-bed SILP carbonylation process design requires a smaller reactor size than existing technology in order to obtain the same productivity, which makes the SILP carbonylation concept potentially interesting for technical applications.

Importantly, in both cases the almost non-volatile nature of the ionic liquid in use and their distinctiveness of being liquid over a large temperature range ensure that the catalyst phase retain on the material in its fluid state even at elevated temperatures. This makes SILP modified materials highly suitable for continuous processing, and further allows the use of traditional, gas-solid fixed-bed reactor technology. In addition, a very large interfacial reaction area is obtained by ensuring a high degree of dispersion of the ionic liquid catalyst phase on the material surface, corresponding to a diffusion layer of only a few molecules of thickness. This provides a very efficient use of the still relatively expensive ionic liquid and circumvents diffusion problems which regularly are found for reactions involving bulk ionic liquids. Thus, the SILP concept allows preservation of high catalyst performance in a technologically simplified and sustainable way.

Additionally, the concept of catalytic SILP materials may be easily combined with several advanced process options providing new opportunities for accomplishing reactions. One attractive approach involves the conductance of consecutive, homogeneous reactions in sequences using several fixed-bed reactors in-series. Another approach involves implementation of integrated reaction-separation techniques using, e.g., SILP-membranes or the use of SILP materials in catalytic distillation processes.

Acknowledgements Financial support for parts of the reported work by the Deutsche Forschungsgemeinschaft (M. Haumann) and by the Danish Research Council for Technology and Production and the Villum Kann Rasmussen Foundation (A. Riisager) is gratefully acknowledged. The Bundesministerium für Bildung und Forschung (BMBF) is acknowledged for generous financial support within the lighthouse project "Regulated Systems for multiphase catalysis/smart solvents – smart ligands".

References

1. Holbrey JD, Seddon KR (1999) Clean Prod Proc 1:223
2. Welton T (1999) Chem Rev 99:2071
3. Dupont J, Consorti CS, Spencer J (2000) J Braz Chem Soc 11:337
4. Sheldon R (2001) Chem Commun 23:2399
5. Olivier-Bourbigou H, Magna L (2002) J Mol Catal A: Chem 182–183:419
6. Stenzel O, Raubenheimer HG, Esterhysen C (2002) J Chem Soc Dalton Trans, p 1132
7. Zhao H, Malhotra SV (2002) Aldrichimica Acta 35:75
8. Welton T (2004) Coord Chem Rev 248:2459
9. Earle MJ, Esperanca JMMS, Gilea MA, Conongia Lopes JN, Rebelo LPN, Magee JW, Seddon KR, Widegren JA (2006) Nature 439(16):831–834
10. Mehnert CP (2005) Chem Eur J 11:50–56
11. Wasserscheid P, Keim W (2000) Angew Chem Int Ed 39:3772
12. Gordon CM (2001) Appl Catal A: General 222:101
13. Dyson PJ (2002) Transition Metal Chem 27:353
14. Zhao D, Wu M, Kou Y, Min E (2002) Catal Today 74:157
15. Dupont J, de Souza RF, Suarez PAZ (2002) Chem Rev 102:3667
16. Wasserscheid P, Waffenschmidt H, Machnitzki P, Kottsieper KW, Stetzler O (2001) Chem Commun 5:451
17. Favre F, Olivier-Bourbigou H, Commereuc D, Saussine L (2001) Chem Commun 15:1360
18. Wasserscheid P, Welton T (eds) (2003) In: Ionic Liquids in Synthesis. Wiley, New York
19. Wasserscheid P, Haumann M (2006) In: Cole-Hamilton D, Tooze B (eds) Catalyst separation, recovery and recycling. Springer, Berlin, Heidelberg, New York. pp 183–213
20. Acros Organics (www.acros.com)
21. Fluka (www.fluka.com)
22. Merck (www.merck.com)
23. Sigma-Aldrich (www.sigma-aldrich.com)
24. Solvent Innovation (www.solvent-innovation.com)
25. Strem (www.strem.com)
26. Wako (www.wako-chem.co.jp)
27. Riisager A, Eriksen KM, Wasserscheid P, Fehrmann R (2003) Catal Lett 90:149–153
28. Riisager A, Wasserscheid P, van Hal R, Fehrmann R (2003) J Catal 219:252–255
29. Riisager A, Fehrmann R, Wasserscheid P, van Hal R (2005) In: Rogers RD, Seddon KR (eds) Ionic liquids IIIB: fundamentals, progress, challenges, and opportunities – transformations and processes. ACS Symposium Series, vol. 902, 23:334
30. Riisager A, Fehrmann R, Flicker S, van Hal R, Haumann M, Wasserscheid P (2005) Angew Chem Int Ed 44:815–819
31. Riisager A, Fehrmann R, Haumann M, Gorle BSF, Wasserscheid P (2005) Ind Eng Chem Res 44(26):9853–9859

32. Riisager A, Fehrmann R, Haumann M, Wasserscheid P (2006) Eur J Inorg Chem, pp 695–706
33. Sherif FG, Shyu L-J (1999) WO9903163, Akzo Nobel Inc., USA
34. deCastro C, Sauvage E, Valkenberg MH, Hölderich WF (2000) J Catal 196:86–94
35. Hölderich WF, Wagner HH, Valkenberg MH (2001) Spec Publ R Soc Chem 266:76–93
36. Valkenberg MH, deCastro C, Hölderich WF (2001) Stud Surf Sci Catal 135:4629–4636
37. Valkenberg MH, deCastro C, Hölderich WF (2001) Top Catal 14:139–144
38. Mehnert CP, Cook RA, Dispenziere NC, Afeworki M (2002) J Am Chem Soc 124:12932
39. Mehnert CP, Mozeleski EJ, Cook RA (2002) Chem Commun, pp 3010–3011
40. Wolfson A, Vankelecom IFJ, Jacobs PA (2003) Tetrahedron Lett 44:1195–1198
41. Hagiwara H, Sugawara Y, Isobe K, Hoshi T, Suzuki T (2004) Org Lett 6:2325–2328
42. Breitenlechner S, Fleck M, Müller TE, Suppan A (2004) J Mol Catal A: Chem 214:175–179
43. van Leeuwen PWNM, Claver C (eds) (2000) In: Rhodium-Catalyzed Hydroformylation. Catalysis by Metal Complexes Series, Kluwer, Dordrecht
44. Young JF, Osborn JA, Jardine FA, Wilkinson G (1965) J Chem Soc Chem Commun, p 131
45. Evans D, Osborn JA, Wilkinson G (1968) J Chem Soc (A), p 3133
46. Evans D, Yagupsky G, Wilkinson G (1968) J Chem Soc (A), p 2660
47. Gregorio G, Montrasi G, Tampieri M, Cavalieri d'Oro P, Pagani G, Andreetta A (1980) Chim Ind (Milan) 62:389
48. Divekar SS, Desphande RM, Chaudhari RV (1993) Catal Lett 21:191
49. Riisager A, Jørgensen B, Wasserscheid P, Fehrmann R (2006) Chem Commun, pp 994–996

Top Organomet Chem (2008) 23: 163–191
DOI 10.1007/3418_2006_058
© Springer-Verlag Berlin Heidelberg
Published online: 16 December 2006

Engineering Aspects of Aqueous Biphasic Catalysis

Yücel Önal · Peter Claus (✉)

Department of Chemistry,
Ernst-Berl-Institute for Technical and Macromolecular Chemistry,
Darmstadt University of Technology, Petersenstr. 20, 64287 Darmstadt, Germany
claus@ct.chemie.tu-darmstadt.de

Abstract Reaction engineering aspects of aqueous multiphase catalysis were investigated in a discontinuously stirred batch reactor as well as in a continuously driven loop reactor. The regioselective hydrogenation of α,β-unsaturated aldehydes to the corresponding unsaturated alcohols with a water-soluble and highly active Ru(II)-TPPTS complex catalyst was chosen as model reaction system. Theoretical and experimental analysis of the mass transport rates at the G/L- and L/L-interphase in the batch reactor ensured that the reaction was limited by kinetics. Recycling experiments with different substrates then revealed that the catalyst phase stability is a function of the water solubility of the reaction products. Taking these aspects into account, kinetic modelling of the reaction rate was performed at elevated hydrogen partial pressure and reaction temperature.

Additional experiments in a loop reactor where a significant mass transport limitation was observed allowed us to investigate the interplay between hydrodynamics and mass transport rates as a function of mixer geometry, the ratio of the volume hold-up of the phases and the flow rate of the catalyst phase. From further kinetic studies on the influence of substrate and catalyst concentrations on the overall reaction rate, the Hatta number was estimated to be 0.3–3, based on film theory.

Keywords Aqueous multiphase catalysis · Regioselective hydrogenation ·
Hydrodynamics · Mass transport · Kinetic modelling

Abbreviations

a_B	G/L interfacial specific surface area [m^{-1}]
a_D	L/L interfacial specific surface area [m^{-1}]
c_i^*	Equilibrium concentration in the catalyst phase [mol L^{-1}]
[CA]	Citral concentration [mol L^{-1}]
D_{aq}	Diffusion coefficient in aqueous phase [m^{2}s^{-1}]
D_{org}	Diffusion coefficient in organic phase [m^{2}s^{-1}]
D_I	Impeller diameter [m]
D_{IO}	Reactor diameter [m]
$d_{32,i}$	Sauter diameter of the dispersed phase i [m]
d_B	Bubble diameter [m]
d_D	Droplet diameter [m]
E_A	Activation energy [kJ mol^{-1}]
G/L	Gas/liquid
g	Acceleration due to gravity [m^{2}s^{-1}]
Ha	Hatta number
k, k_1	Reaction rate constant [L mol^{-1}bar^{-1}min^{-1}]
k_2	Inhibition constant [L mol^{-1}]
k_L	G/L overall mass transport coefficient [min^{-1}]
k_{LL}	L/L overall mass transport coefficient [min^{-1}]
K_H	Henry coefficient [mol L^{-1}bar^{-1}]
K_N	Nernst coefficient
L/L	Liquid/liquid
L	$= D_I/D_{IO}$
N	Stirring rate [min^{-1}]
P_{H2}	hydrogen partial pressure [bar]
[PA]	Prenal concentration [mol L^{-1}]
[PO]	Prenol concentration [mol L^{-1}]
r	Reaction rate [mol L^{-1}min^{-1}]
r_{calc}	Calculated reaction rate [mol L^{-1}min^{-1}]
r_{exp}	Experimentally obtained reaction rate [mol L^{-1}min^{-1}]
[Ru]	Ruthenium concentration [mol L^{-1}]
Re	Reynolds number
Sh	Sherwood number
Sc	Schmidt number
T	Temperature [K]
T_R	Temperature [°C]
TPP	Triphenylphosphane
TPPTS	Triphenylphosphane-33′3″-trisulfonate-Na-salt
V_{aq} (Batch reactor)	Reaction volume – catalyst phase [mL]
V_{org} (Batch reactor)	Reaction volume – organic phase [mL]
V_{aq} (Loop reactor)	Volumetric flow rate – catalyst phase [mL min^{-1}]
V_{org}(Loop reactor)	Volumetric flow rate – organic phase [mL min^{-1}]
V_{Feed} (Loop reactor)	Volumetric flow rate – feed [mL min^{-1}]
V_{H2} (Loop reactor)	Volumetric flow rate – hydrogen [mL min^{-1}]
$V_b(B)$	Molar volume at boiling point (solvent) [mL mol^{-1}]

$V_b(A)$	Molar volume at boiling point (solute) [mL mol^{-1}]
We	Weber number
α	Reaction order – aldehyde concentration
β	Reaction order – hydrogen partial pressure
β_{gas}	Mass transport coefficient related to the gas phase [min^{-1}]
β_{aq}	Mass transport coefficient related to the aq. phase [min^{-1}]
β_{org}	Mass transport coefficient related to the org. phase [min^{-1}]
χ	Reaction order – Ruthenium concentration
δ_{aq}	Aqueous boundary layer length [m]
ε_{org}	Volume hold-up – organic phase
ε_{aq}	Volume hold-up – aqueous phase
η_B	Dynamic viscosity of solvent [cP]
φ_{aq}	Density of water [kg m^{-3}]
$\Delta\varphi$	Density difference of aqueous and organic phase [kg m^{-3}]
μ_{aq}	Aqueous phase viscosity [kg m^{-1}s^{-1}]
σ_B	Surface tension of solvent [mN m^{-1}]
σ_A	Surface tension of solute [mN m^{-1}]
Ψ	Enhancement factor

1
Introduction

Homogeneous catalysts are highly selective and active in the synthesis of fine chemicals at an industrial scale. In comparison to heterogeneous catalysts, mild reaction conditions are applied. Moreover, the selectivity of the desired reaction can be fine-tuned by a special ligand design. One major drawback of homogeneous catalytic processes is the high costs of metal precursors and of ligand synthesis. In order to reuse the catalyst the plant design, including for example distillation columns for catalyst recovery, is very complex. Therefore, a homogeneous catalysed process is mostly only feasible in the synthesis of fine chemicals, whereas bulk chemicals and commodities are mainly produced by heterogeneous catalysis.

By introducing the principles of aqueous multiphase catalysis for the hydroformylation of short chain olefins to aldehydes, Ruhrchemie/Rhône Poulenc established a facile catalyst separation approach, providing a more economical way of process design in homogeneous catalysis [1–6]. The reaction is carried out in a multiphase system comprising a polar aqueous phase with the water-soluble catalyst and a non-polar solvent phase solubilizing the educts and products of the desired reaction. The solubility of the catalysts in the aqueous phase is provided by highly water-soluble ligands, which are prepared from common phosphine ligands by structural modification with polar functional groups ($-SO_3^-$, $-OH$, NR_3^+, $-COOH$). Triphenylphosphane-3,3′,3″-trisulfonate-Na-salt (TPPTS), which is also used on an industrial scale and exhibits a water solubility of $1100\,g\,L^{-1}$ [6], was used as ligand in our investigations.

Besides the advantages of process design, special emphasis is also put on water, which is an environmentally benign solvent for industrial purposes.

During reaction the educts have to pass the phase boundary at the G/L- or L/L-interphase. The reaction is then carried out in the aqueous homogeneous phase, followed by fast extraction of the products back to the organic phase. The catalyst phase can be easily recovered by phase separation and reused without further purification. In order to reach high space–time yields, the educts must be soluble to a certain degree in the aqueous phase and fast mass transport velocities must be provided either by high stirring rates in a batch reactor or high volume rates in a tube reactor. In most cases, the organic substrate phase as well as the gas phase, if present, are dispersed in the continuous aqueous phase, so that mass transport at the G/L- and L/L-phase boundaries and the intrinsic kinetic rates have to be evaluated simultaneously in order to develop a realistic kinetic model. Understanding reaction engineering aspects in aqueous multiphase catalysis is therefore very complex and a number of physico-chemical features of the reaction system have to be evaluated. These are summarized in Fig. 1. Review articles on this topic have been very recently published by Chaudhari and Claus et al. [7, 8].

Intrinsic kinetic data can only be measured provided that the overall reaction rate is not limited by mass transport. Only then realistic parameters can be calculated concerning the influence of catalyst and substrate concentrations (reaction order) as well as the temperature dependency (activation

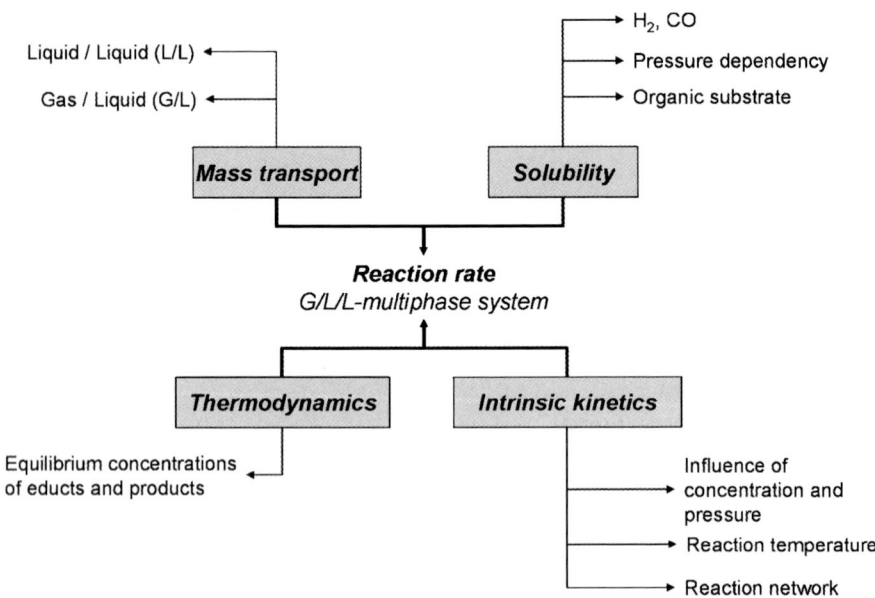

Fig. 1 Complex interplay of physico-chemical and kinetic aspects in aqueous multiphase catalysis

energy) on the reaction rate. Thermodynamic equilibria, e.g. gas solubility and distribution coefficients of educt and product, can be directly implemented in the mathematic rate model. If the reaction is carried out in a mass transport limited regime, thermodynamic equilibria would not be reached and the overall reaction rate would be a result of a complex interplay of mass transport and reaction. Another point that also determines the kinetic model is the reaction site, as demonstrated by Cornils et al. [9]. If the reaction took place at the phase boundary, the substrates would not have to pass the boundary layer by diffusion (Fig. 2; Case 1: chemical reaction in the bulk of the aqueous phase; Case 2: chemical reaction at the L/L-interphase). Mass transport terms in the kinetic model and also equilibrium concentrations have then to be modified.

Important hints on the reaction site can be gained by the Hatta numbers (Ha) of mass transport at the G/L- and L/L-phase boundaries. These numbers are also essential in order to estimate mass transport rates and concentration profiles within the boundary layer. Since the main resistance of mass transport is in the aqueous phase, mass transport coefficients and Ha numbers mentioned in the text are related to the aqueous phase.

The main objective of the present work was to gain a better insight and understanding of mass transport and reaction interactions in aqueous multiphase catalysis by an innovative reactor design. Reaction engineering aspects were investigated in a discontinuously working batch reactor as well as in a continuously driven tube reactor with static mixers. Experiments carried out in the tube reactor also allowed a better understanding of the influence of hydrodynamics (volume rate, bubble/droplet size) on mass transport and reaction rate.

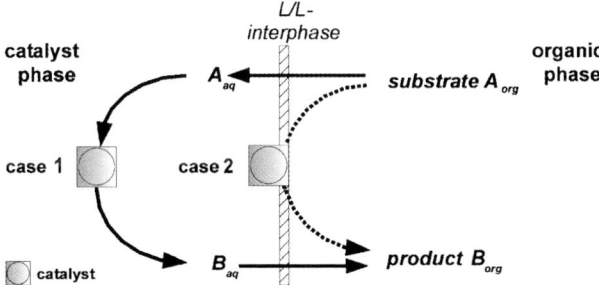

Fig. 2 Location of reaction within the bulk of the catalyst phase or at the L/L-phase boundary

2
Experimental Setup

2.1
Model Reaction System

All kinetic investigations carried out in this work are based on the regioselective hydrogenation of unsaturated aldehydes to unsaturated alcohols, which are important compounds in the synthesis of fine chemicals and fragrances. The aldehyde exhibits two reactive sites, namely the $C = C$ and the $C = O$ bond, for hydrogenation. Thereby the hydrogenation of the $C = C$ bond is thermodynamically and kinetically favoured and a complex reaction network has to be taken into account (Fig. 3). Nevertheless, high reaction rate and selectivity to the desired product are obtained with a water-soluble Ru(II)-TPPTS complex catalyst, which is generated in situ from the precursor $RuCl_3 * 3H_2O$ and TPPTS at relatively mild reaction conditions.

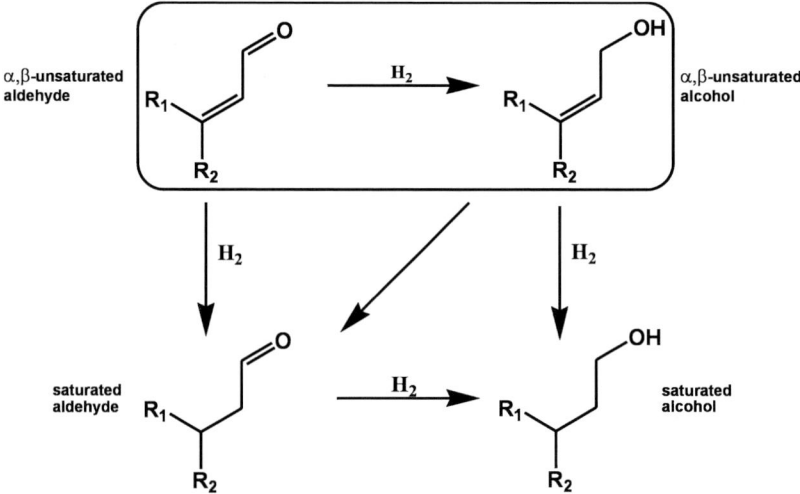

Fig. 3 Reaction network in the hydrogenation of α,β-unsaturated aldehydes

Chemical aspects concerning the mechanism of catalyst formation and reaction have already been published by Grosselin and Joó et al. [10, 11]. According to results published by Joó et al. on the Ru(II)-TPPMS complex, the pH of the aqueous phase has an important effect on the chemical structure of the catalyst formed and hence the selectivity in the hydrogenation reaction. Thereby high selectivity to the unsaturated alcohol was obtained in the hydrogenation of cinnamaldehyde at pH values higher than 8, which was based on NMR studies attributed to the formation of $[H_2Ru(TPPMS)_4]$ (Fig. 4).

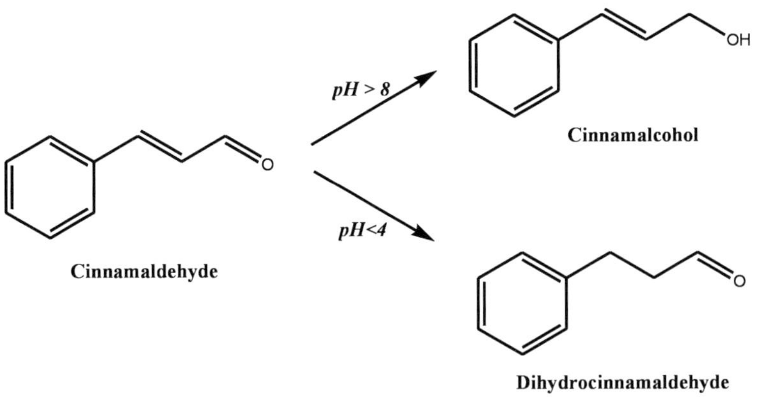

Fig. 4 Regioselectivity in the hydrogenation of cinnamaldehyde with Ru(II)-TPPMS as a function of pH

Furthermore, the amount of ligand present in the reaction system must be higher than expected from stoichiometry, since the ligand is partly oxidized during the formation of the catalyst [10]. The main substrates investigated in this work were prenal (3-methylcrotonaldehyde) and citral (Fig. 5), although the transferability of the catalyst system to other α,β-unsaturated aldehydes was also checked.

Fig. 5 Chemical structures of prenal and citral

One of the physical features that is totally different for prenal and citral is their water solubility. In contrast to prenal ($110\,\mathrm{g\,L^{-1}}$), citral is barely soluble in water ($1.0\,\mathrm{g\,L^{-1}}$). Since the reaction is supposed to take place in the bulk of the aqueous phase or in the boundary layer, this aspect gives a first hint about whether the reaction could be limited by mass transport or kinetics.

2.2
Batch Reactor

Preliminary experiments to optimize the reaction conditions with respect to selectivity and activity (in situ catalyst generation, pH value, [P]:[Ru] ratio)

as well as the kinetics of the hydrogenation reaction were first carried in a batch reactor (Parr Instruments). Reactions were performed at elevated hydrogen partial pressure ($<$ 40 bar) and reaction temperature ($<$ 70 °C). The total volume of the aqueous and organic phases added up to 150 mL, whereas the volume ratios of both phases were the same (75 mL). The reactor was equipped with an electronic heating jacket and a mass flow controller connected to a high pressure gas reservoir, so that the reaction could be carried out isobarically ($\pm$0.1 bar). The temperature as well as the pressure within the reactor was controlled by an electronic control unit connected to a PC. Simultaneously, the time-dependent gas uptake volume representing the course of the reaction was also recorded. Stirring was provided by a specially designed gas uptake stirrer, which was magnetically coupled to a rotor. The stirring rate ($<$ 2000 rpm) could also be adjusted by the control unit. The inner wall of the reactor was covered with Teflon, so that a catalytic effect of the reactor material could be neglected. Furthermore, the reactor was equipped with a sampling system so that concentration–time profiles could be measured simultaneously by GC analysis.

RuCl$_3$*3H$_2$O (0.005 M) and TPPTS (0.05 M) were dissolved in an aqueous buffered (pH 7) solution, whereas the aldehyde (0.5 M) was prepared as a solution in n-hexane or toluene. Prior to reaction the aqueous and organic phases were added to the reactor and mixed at moderate stirring rates 300–400 rpm). The reaction mixture was first flushed three times with argon and then with hydrogen. Then the mixture was heated up to the desired temperature (2–3 K min^{-1}) and afterwards pressurized with hydrogen. The reaction was started by accelerating the stirring rate (1500–2000 rpm).

2.3
Loop Reactor

A schematic representation of the loop reactor is shown in Fig. 6. The reaction is carried out in a tube reactor (V_r = 61 mL) equipped with static mixers. The three phases, namely the aqueous, organic and hydrogen phases, are pumped into the reactor by circulating pumps or a mass flow controller, respectively. The volume ratio of the catalyst phase to the organic phase is typically 6 to 1, so that the organic as well as the gas phase are dispersed in the continuous catalyst phase. Within the phase separator, the reaction mixture is then separated into the aqueous catalyst phase at the bottom and the organic phase with the lower density above. The catalyst phase is fully recycled to the reactor, whereas the organic phase is partly (6%) withdrawn from the reaction system. The same amount of educt stream is fed into the reactor by a HPLC pump. The product phase leaving the reactor passes a 6-port valve connected to an online GC, so that the course of reaction can be followed simultaneously. Redundant hydrogen is removed from the top of the phase separator by a venting system equipped with a further mass flow controller. Due to the

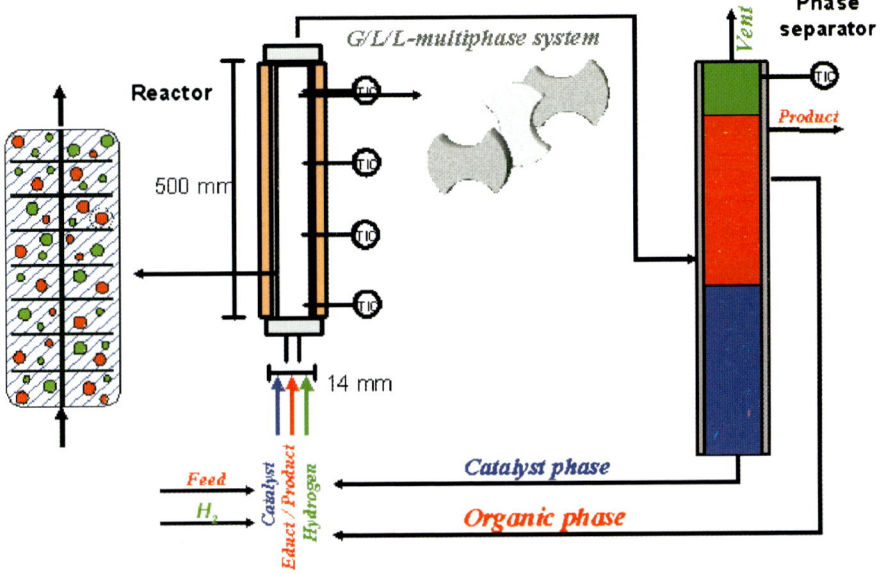

Fig. 6 Schematic representation of the loop reactor

continuous dosing of hydrogen into the reactor, the partial pressure of hydrogen would increase with reaction time. In order to avoid this and to carry out the reaction isobarically, hydrogen is vented from the reactor. This happens automatically by a PC connected to the control unit of the mass flow controller and to the electronic pressure device. Heating of the reactor is provided by water, whereas the phase separator is heated electronically. Both units are kept at the same temperature, so that cooling of the reaction mixture within the phase separator could be neglected. The residence time of the organic phase within the separator is higher then in the reactor. This is due to the phase separation, which needs a certain time (at least about 10 s) for a clean coagulation. Some important features of the loop reactor can be seen in Table 1.

All reaction parameters including volume rates of the hydrogen, organic and aqueous phase as well as the partial pressure of hydrogen and the reaction temperature are recorded on a PC and can be controlled by process control software, which was developed in the working group.

Table 1 Specifications of the loop reactor

p [bar]	1–40	V_{H2} [mL min^{-1}]	0–1000
T [°C]	RT–80	V_{Feed} [mL min^{-1}]	0–10
V_{aq} [mL min^{-1}]	0–2000	V_{org} [mL min^{-1}]	0–250

Only the hydrogenation of citral with Ru(II)-TPPTS was carried out in the loop reactor, because no deactivation was observed in this case (see Sect. 3.1.2). The phase separator was filled with 200 mL aqueous catalyst phase comprising $RuCl_3*3H_2O$ (0.005 M) and TPPTS (0.05 M) in a buffered solution (pH 7). The circulating pump was switched on (300 mL min^{-1}) and the catalyst phase was first flushed three times with argon and then with hydrogen. Afterwards, the reactor was pressurized with hydrogen (5 bar) and the catalyst phase was pre-treated for 1 h at 60 °C and a hydrogen volume rate of 70 mL min^{-1}. Then the reactor was depressurized and the organic phase (300 mL) with citral (0.25 M) was added to the reactor. The reaction was again put under hydrogen pressure and the circulating pump of the organic phase switched on (50 mL min^{-1}). Typically, reactions were carried out at a hydrogen volume rate of 100 mL min^{-1} and a citral-feed rate of 3 mL min^{-1}. During reaction, the concentration profiles of the educt and product were measured by an online GC. After roughly 2 h, the loop reactor reached stationary reaction conditions. The reaction was then followed for about 2 h further at the same reaction conditions.

Measurements of the bubble and droplet sizes were performed during reaction in a glass tube reactor by photographic imaging with a fast resolution time. Thereby, reactions could be carried out at maximal 5 bar. The images were taken with a digital photo camera (Casio QV-4000) and then statistically analysed on a PC by the software *Lince* (Freeware from the Material Sciences Department, Darmstadt University of Technology) concerning bubble and droplet size distributions.

3
Results and Discussion

3.1
Batch Reactor

3.1.1
Hydrogenation of α,β-Unsaturated Aldehydes with Ru(II)-TPPTS

After optimization of reaction conditions with a special focus on in situ catalyst generation, the pH value of the catalyst phase and the ratio of ligand to metal in the hydrogenation of prenal, the transferability of the catalyst system to other α,β-unsaturated aldehydes was checked. The influence of steric hindrance at the C3-atom and the water solubility of the substrates on the reaction rate and selectivity to the unsaturated alcohol were analysed (Table 2). The initial concentration of the aldehyde in the organic phase was always 0.5 M. Apart from acrolein, which is not mentioned in the table, generally all kinds of α,β-unsaturated aldehydes can be selectively hydrogenated with

Table 2 Hydrogenation of different α,β-unsaturated aldehydes at optimized reaction conditions

Aldehyde	Solvent	Water solubility [g/L]	t [min]	X [%]	S [%]
Crotonaldehyde	$H_2O/$ Toluene	181	40	99	90
Prenal	$H_2O/$ n-hexane	110	40	99	99
Cinnamaldehyde	$H_2O/$ toluene	1.5	100	99	99
Citral	$H_2O/$ n-hexane	1.0	180	94	99

Reaction conditions: $T = 50\,°C$, $p_{H2} = 20\,bar$, [Ru] = 0.005 M, [P] = 0.05 M, [Aldehyde] = 0.5 M

Ru(II)-TPPTS to the corresponding unsaturated alcohols in biphasic mode. If one compares the reaction times until full conversion, it becomes clear that the reaction rate correlates with the solubility of the substrate in the aqueous phase, as expected. The latter decreases with increasing chain length or branching of the chain at the C3-atom. In contrast to heterogeneously catalysed hydrogenations of α,β-unsaturated aldehydes, the steric hindrance of substituents at the C3-atom only plays a minor role in the coordination mode of the substrate at the metal centre, since selectivity differences from crotonaldehyde to citral are marginal.

3.1.2
Recycling Experiments

Prior to the kinetic experiments, possible deactivation phenomena of the catalytic system were checked by recycling experiments with prenal and citral as substrates. These results provide not only important hints on the form of the rate equation, but also on which reaction is convenient for long-term investigations in the loop reactor. After the reaction, the aqueous and organic phases were separated and the catalyst phase was reused without further purification. Results on the hydrogenation of prenal are shown in Fig. 7. The reaction rate clearly decreases if the catalyst phase is reused. According to GC analysis and ^{1}H-NMR studies, this can be attributed to the fact that the product of the reaction, prenol, is highly soluble in water. Consequently, a simple phase

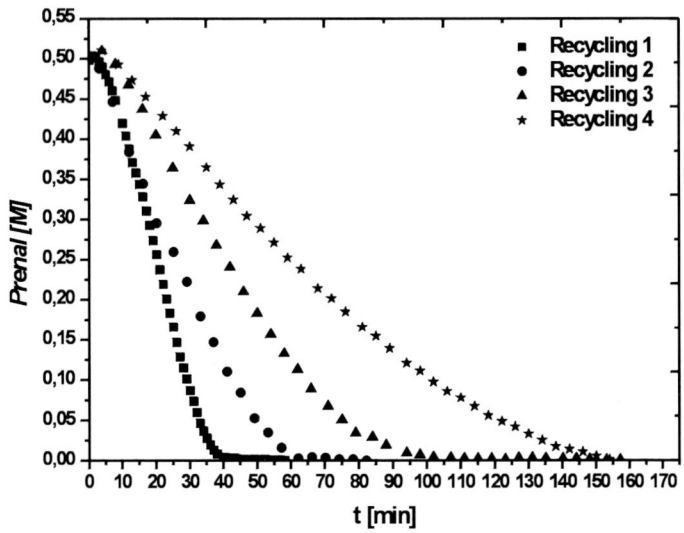

Fig. 7 Recycling experiments with prenal as substrate; $p_{H2} = 20$ bar, $T = 50\,°C$, [Prenal] = 0.5 M, [Ru] = 0.005 M, [TPPTS] = 0.05 M, $V_{aq} = V_{org} = 75$ mL, solvent was *n*-hexane

separation without an extraction step is not enough to remove prenol quantitatively from the aqueous phase. As a result, reactive sites of the complex remain blocked by the prenol and the activity of the catalyst phase decreases. NMR studies showed that catalyst deactivation cannot be attributed to leaching of the complex into the organic phase. A similar observation concerning hydrogenation of prenal, based on supported aqueous phase catalysts (SAPC) with Ru(II)-TPPTS, was also reported by Fache et al. [12]. Again, catalyst deactivation was attributed to an accumulation of products in the aqueous catalyst layer. Due to the deactivation of the catalyst phase, it is obvious that an inhibiting term has to be taken into account in order to describe the reaction rate mathematically. The hydrogenation of prenal is therefore not a practical model reaction for kinetic investigations in the loop reactor. The overall reaction rate would decrease and kinetic data obtained would not be representative, since the catalyst phase would deactivate with reaction time.

In contrast to the results obtained in the hydrogenation of prenal, no deactivation was observed for citral hydrogenation in recycling experiments carried out with the same aqueous catalyst phase (Fig. 8). Moreover, the reaction rate increases after the first run, which can be attributed to the fact that the catalyst generated in situ is still active in the following runs and needs no induction period for activation. Unlike the results obtained with prenal, this behaviour can be explained in terms of advantageous distribution coefficients for nerol and geraniol, which are barely soluble in water. Due to this fact, only the hydrogenation of citral was chosen as a model reaction for long-term kinetic experiments in the loop reactor.

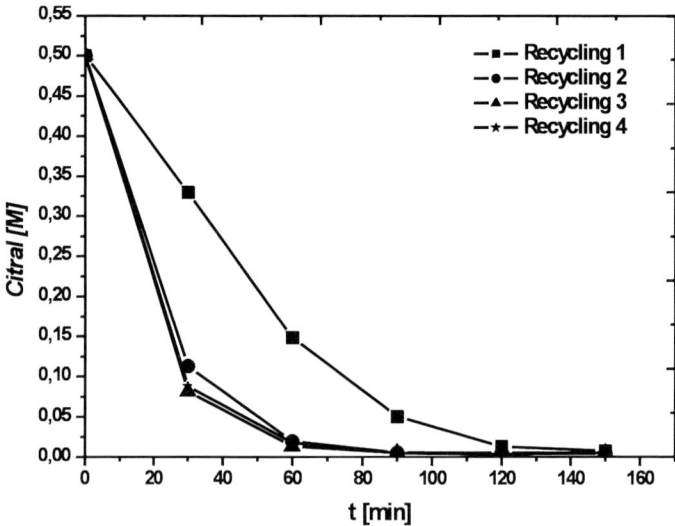

Fig. 8 Recycling experiments with citral as substrate; $p_{H2} = 20$ bar, $T = 60\,°C$, [Prenal] = 0.5 M, [Ru] = 0.005 M, [TPPTS] = 0.05 M, $V_{aq} = V_{org} = 75$ mL, solvent was n-hexane

3.1.3
Evaluation of Mass Transport Rates

Reactions carried in aqueous multiphase catalysis are accompanied by mass transport steps at the L/L- as well as at the G/L-interface followed by chemical reaction, presumably within the bulk of the catalyst phase. Therefore an evaluation of mass transport rates in relation to the reaction rate is an essential task in order to gain a realistic mathematic expression for the overall reaction rate. Since the volume hold-ups of the liquid phases are the same and water exhibits a higher surface tension, it is obvious that the organic and gas phases are dispersed in the aqueous phase. In terms of the film model there are laminar boundary layers on both sides of an interphase where transport of the substrates takes place due to concentration gradients by diffusion. The overall transport coefficient k_L can then be calculated based on the resistances on both sides of the interphase (Eq. 1):

$$\frac{1}{(k_L a)} = \frac{1}{\psi_i \, (\beta_{aq} a)_i} + \frac{K_H}{(\beta_G a)_i}, \tag{1}$$

where β is the transport coefficient of a particular phase, K_H the Henry coefficient for the solubility of hydrogen in water, ψ_I the enhancement factor and a the interfacial specific surface area. Mass transport at the L/L-phase boundary can be expressed analogously by substituting K_H with the Nernst distribution coefficient K_N, and β_G with the transport coefficient in the organic boundary layer. Provided that the transport coefficients in the gas and organic phase are large compared to the aqueous boundary layer, the second term in Eq. 1 can be neglected and the overall transport coefficient can just be related to the aqueous phase. Evaluating mass transport rates can therefore be simplified to the determination of $k_L a$ (G/L) and $k_{LL} a$ (L/L) values. This can be done experimentally as described elsewhere [13] or by theoretical calculations [14, 15]. Values of $k_L a$ for the rate of gas uptake as a function of the solvent and stirring rate were determined by the dynamic absorption method described in [13]. Results are shown in Table 3.

In order to evaluate k_{LL}, the diffusion coefficients of prenal and citral were calculated based on the Tyn–Calus method (Eq. 2):

$$D_{AB} = 8.93 \times 10^{-8} \frac{V_{b(B)}^{0.267}}{V_{b(A)}^{0.433}} \frac{T}{\eta_B} \left(\frac{\sigma_B}{\sigma_A}\right)^{0.15} \left[\frac{cm^2}{s}\right]. \tag{2}$$

The resulting values are shown in Table 4. As expected, the diffusion coefficients of prenal and citral are smaller in water than in n-hexane. Since the mass transport coefficients in each boundary layer directly correlate with the diffusion coefficient (Eq. 3), this result confirms the assumption that the overall mass transport resistance can be predominantly referred to the aqueous catalyst phase:

$$k = \frac{D}{\delta}. \tag{3}$$

Table 3 $k_L a$ values obtained for different solvents as a function of the stirring rate

Stirring rate	$k_L a$			
	n-Hexane	Toluene	Ethyl acetate	Aq. buffered solution
[rpm]	[min^{-1}]	[min^{-1}]	[min^{-1}]	[min^{-1}]
230	0.09	0.02	0.15	0.03
641	0.45	0.17	0.80	0.13
1110	5.82	2.63	2.69	2.95
1542	12.38	4.08	–	7.26
1932	15.86	4.85	5.39	11.4

Table 4 Calculated diffusion coefficients for prenal and citral in n-hexane and water at 25 °C

	D_{aq} in H_2O [m^{2}s^{-1}]	D_{org} in n-hexane [m^{2}s^{-1}]
Citral	7.3×10^{-10}	9.9×10^{-10}
Prenal	8.96×10^{-10}	1.20×10^{-9}

The overall mass transport coefficient for mass transport from the dispersed organic phase into the continuous aqueous phase was then calculated according to the Calderbank equation (Eq. 4):

$$k_{LL} = 0.31 \left(\frac{\Delta \varphi \mu_{aq} g}{\varphi_{aq}^2} \right)^{1/3} \left(\frac{\mu_{aq}}{\varphi_{aq} D_{aq}} \right)^{-2/3}. \tag{4}$$

In order to estimate the specific surface area of the dispersed organic droplets, the mean droplet size (Sauter diameter d_{32}) has to be determined, which can be calculated according to the Okufi equation (Eq. 5):

$$d_{32,org} = 0.126 \left(1 + 2\varepsilon_{org} \right) We^{-0.6} L^{-0.4} D_I \, [m] \tag{5}$$

$$We = \frac{\varphi_{aq} N D_I}{\sigma_{aq}}.$$

The specific surface area can then be determined based on Eq. 6:

$$a_D = \frac{6\varepsilon_{org}}{d_{32,org}}. \tag{6}$$

$k_{LL}a$ values for prenal and citral obtained by this approach are listed in Table 5.

With the knowledge of the equilibrium concentrations of hydrogen and aldehyde in water at reaction conditions, the maximum mass transport rates can be determined, assuming that the concentration of the substrate in the aqueous phase is zero (Eqs. 7 and 8):

$$J_{H_2} = k_L a_B c^*_{H_2,aq} \tag{7}$$

$$J_{Aldehyde} = k_{LL} a_D c^*_{Aldehyde,aq}. \tag{8}$$

In order to evaluate whether a mass transport limitation has to be taken into account or the reaction is limited by kinetics, the observed reaction rate r has to be set in relation to the maximum mass transport rates J_i (Table 6).

The ratio of the rate of intrinsic kinetics to mass transport at the L/L-interphase is expressed by the Ha number (Eq. 9). According to Chaudhari et al. [14], the Ha numbers are smaller than 0.3 as long as the ratio of reaction rate to mass transport rate are not higher than 0.1. It is therefore concluded

Table 5 Overall mass transport coefficients k_{LL} and $k_{LL}a$ values for the mass transport of prenal and citral in the biphasic system n-hexane/water

	Prenal	Citral
k_{LL} [m s^{-1}]	0.124×10^{-3}	0.108×10^{-3}
$k_{LL}a_D$ [s^{-1}] ($a = 13\,043$ m^{-1})	1.62	1.41

Table 6 Ratios of observed reaction rate to the calculated mass transport rates at G/L- and L/L-interphase

	$\dfrac{r}{k_L a_B c^*_{H_2}}$	$\dfrac{r}{k_{LL} a_D c^*_{Aldehyde}}$
Hydrogen	0.07	–
Prenal	–	0.0025
Citral	–	0.029

that the overall reaction rates observed in the present batch experiments are limited by kinetics:

$$\text{Ha} = \delta_{aq}\sqrt{\frac{k\,(T)}{D_{aq}}}\,. \tag{9}$$

Another point that can be evidenced by the Ha number is the location of reaction. Since the Ha numbers are very small regarding mass transport at the G/L- as well as at the L/L-interphase, the reaction takes place predominantly in the bulk of the catalyst phase. Experimental work on the influence of the stirring rate on the overall reaction rate also underlines the results obtained by theoretical calculations. There were no detectable changes in the reaction rate as long as stirring rates > 1000 rpm were applied. Therefore, kinetic experiments were typically performed at 2000 rpm.

3.1.4
Kinetic Modelling

Kinetic experiments on the hydrogenation of prenal and citral were first carried out in the batch reactor with variation of aldehyde concentration, hydrogen partial pressure, ruthenium concentration and reaction temperature (Table 7). Stirring was provided at 2000 rpm in order to be sure that the overall reaction rate was determined by kinetics.

Kinetic modelling was then performed, based on the obtained concentration–time profiles assuming semi-empirical rate equations (Eqs. 10, 11):

Hydrogenation of prenal

$$r = -\frac{d[PA]}{dt} = \frac{k_1 [PA]^\alpha_{org} p^\beta_{H_2,g} [Ru]^\chi_{aq}}{\left(1 + k_2 [PO]_{aq}\right)} \tag{10}$$

Hydrogenation of citral

$$r = -\frac{d\,[CA]}{dt} = k\,[CA]^\alpha_{org}\, p^\beta_{H_2}\,[Ru]^\chi_{aq}\,. \tag{11}$$

Table 7 Reaction conditions for kinetic modelling of prenal and citral hydrogenation

Parameter	Prenal hydrogenation interval	Citral hydrogenation interval
[Ru]	2.5–10 mmol L^{-1}	2.5–7.5 mmol L^{-1}
[PA], [CA]	0.25–1 mol L^{-1}	0.5–1 mol L^{-1}
p_{H2}	10–40 bar	5–20 bar
T_R	30–70 °C	50–80 °C

Deactivation of the catalyst phase due to the fact that prenol is not fully with-drawn from the aqueous phase is taken into account by an inhibiting term in the denominator of the rate equation. Since a significant deactivation was

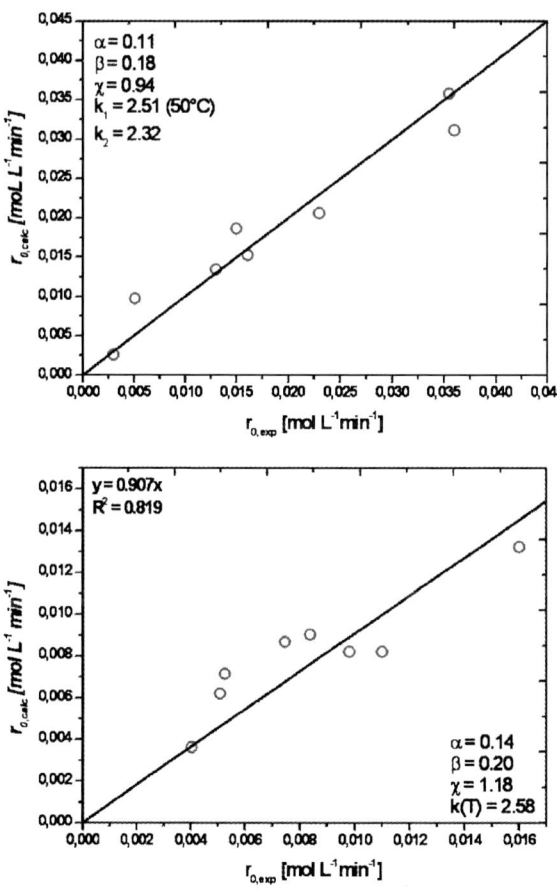

Fig. 9 Correlation of theoretical and experimental reaction rates for prenal (*top*) and citral hydrogenation (*bottom*)

not observed for citral hydrogenation, a simple power law can be used to describe the reaction rate. The kinetic parameters were then calculated using an optimization routine implemented in the software *Presto Kinetics*. Correlation diagrams of the calculated and experimentally obtained reaction rates are shown in Fig. 9.

Interestingly, the reaction orders for hydrogen partial pressure as well as the aldehyde concentration are almost zero. The reaction rate seems to be only affected by the concentration of ruthenium in the aqueous phase and by the reaction temperature. The rate constants k_1 (prenal hydrogenation) and k (citral hydrogenation) were analysed according to an Arrhenius plot. Activation energies E_A were found to be $54\,\text{kJ}\,\text{mol}^{-1}$ with respect to prenal hydrogenation and $34\,\text{kJ}\,\text{mol}^{-1}$ with respect to citral hydrogenation.

3.2
Loop Reactor

3.2.1
Static Mixers and Bubble/Droplet Sizes

The influence of hydrodynamics on the reaction rate was analysed based on the results obtained in the loop reactor. Again only the hydrogenation of citral was investigated as a model reaction, since no deactivation was observed over a reaction period of 4 h. The solubility of citral in water is very low and an effective mixing of the biphasic system has to be provided in order to reach good space–time yields. The efficiency of different static mixers and of carrier materials was analysed initially in order to check the mixing efficiency on the bubble/droplet diameters, on hydrodynamics, and hence on the overall reaction rate. Hydrogenation was carried out at 5 bar in a glass tube, so that the droplets and bubbles formed could be analysed visually by a digital camera. Images of the differently shaped static mixers and the carrier materials used are shown in Fig. 10. Overall reaction rates obtained with the mixers and carrier materials in comparison to the reaction rates obtained without mixing are depicted in Fig. 11.

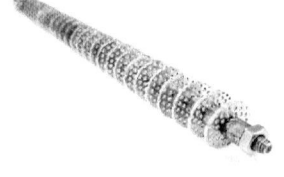  

Fig. 10 Different static mixers and carrier materials applied for our kinetic investigations in the loop reactor; *left* sieve plates, *centre* double axe, *right* carrier materials (twisted)

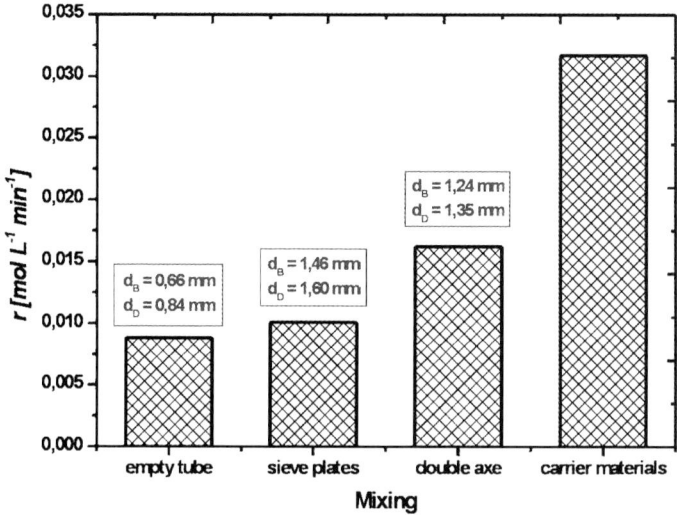

Fig. 11 Overall reaction rates obtained with static mixers and carrier materials in comparison to the reaction rate without mixing; $T = 50\,^\circ$C, $p_{H2} = 5$ bar, $V_{aq} = 300$ mL min^{-1}, $V_{org} = 50$ mL min^{-1}, $V_{Feed} = 3.0$ mL min^{-1}, $V_{H2} = 100$ mL min^{-1}

Highest reaction rates were obtained with the carrier materials, whereas the efficiency of the static mixers is clearly lower. The efficiency of a mixer is based on two parameters: firstly, the formation of a high specific surface area by generation of small bubbles d_B and droplets d_T and, secondly, the occurrence of high turbulence, which enhances mass transport rates. Unfortunately, the droplet and bubble diameters with the carrier materials could not be derived visually. Moreover, implementing carrier materials in a reactor for mixing is coupled with high operational costs, since high pressure drops have to be taken into account. However, the bubble and droplet sizes could be measured accurately using the static mixers. Together with the calculated specific surface areas these values are listed in Table 8 in comparison with values obtained in the empty tube.

Table 8 Mean bubble and droplet diameters and resulting specific surface areas of the dispersed gas and organic phase

	Empty tube	Double axe	Sieve plates
d_B	0.66 mm	1.24 mm	1.46 mm
a_B	491 m^{-1}	261 m^{-1}	222 m^{-1}
d_D	0.84 mm	1.35 mm	1.60 mm
a_D	1007 m^{-1}	627 m^{-1}	529 m^{-1}

In the empty tube, bubble and droplet sizes are clearly smaller and hence specific surface areas at the G/L- and L/L-interphase are higher than with the static mixers. Obviously, contact of the dispersed phases with the mixer plates supports the coagulation of bubbles and droplets. However, the overall reaction

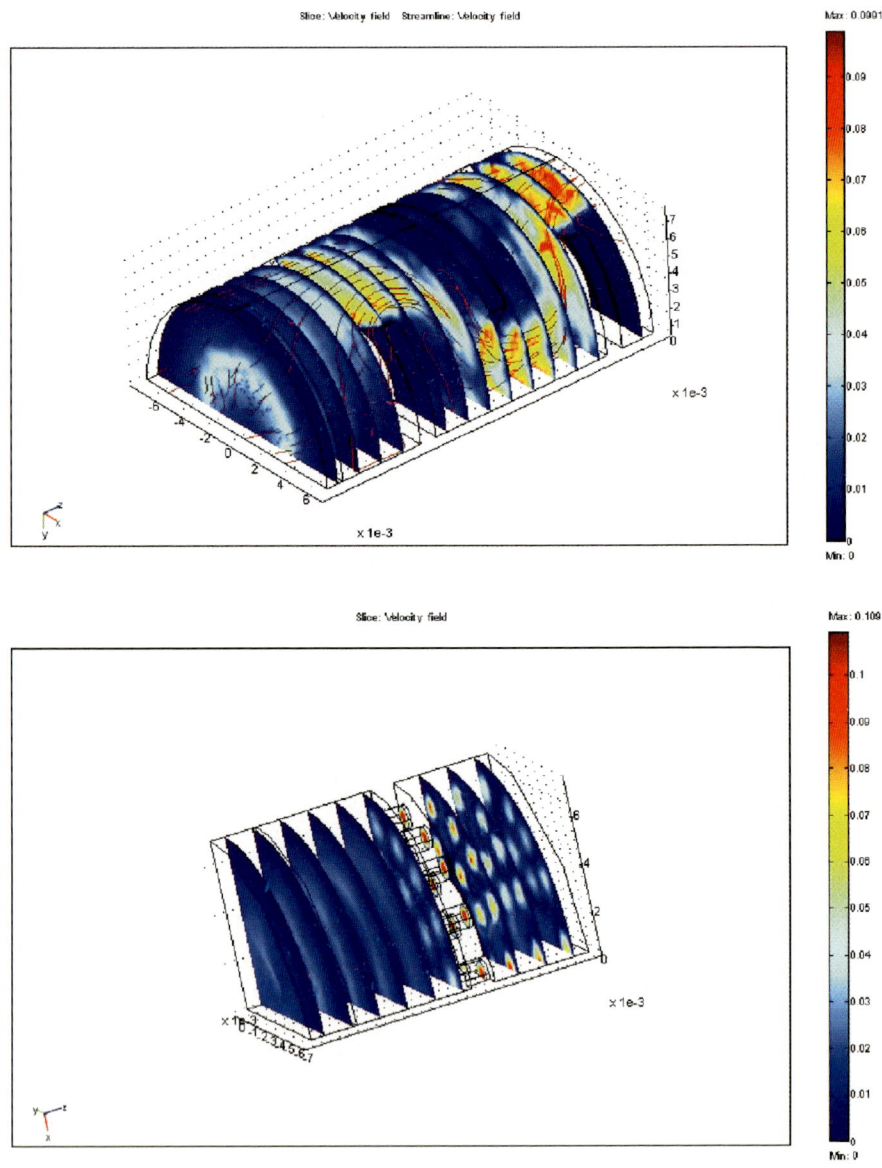

Fig. 12 CFD Images of the flow pattern as a function of the static mixer geometry for a periodic section of the reactor; *top* double axe, *bottom* sieve plates

rate in the empty tube is slower than with the static mixers, for which overall reaction rates correlate with the specific surface areas, as expected. As mentioned, besides the formation of high specific surface area, hydrodynamics in the continuous phase also play a decisive role on the rate of mass transport. Typical Reynolds (Re) numbers regarding the continuous phase in the empty tube are around 980, indicating laminar flow. As a matter of fact, the mass transport coefficients are small and the resulting overall reaction rates are slower than with the static mixers. Turbulence within the continuous phase is generated by using static mixers, enhancing the mass transport. Although static mixers provide smaller surface areas this turbulence accelerate mass transport and hence the reaction rate. In order to gain a better insight into hydrodynamics when static mixers are used, the flow patterns of the continuous phase within the specific geometry of the tube reactor were simulated using CFD software (*Femlab 3.0*). Images of the flow for a small periodic section of the reactor are shown in Fig. 12 for the static mixers sieve plates and double axe.

Although the inlet feed is originally laminar-like in the empty tube, using static mixers local stream velocities can be accelerated by a factor of 10, supporting mass transport rates. Nevertheless, the flow pattern generated with the static mixers double axe and sieve plates are not equal, giving rise to the experimentally observed differences in the overall reaction rates. By using the static mixer double axe, turbulent areas are generated whenever the flow is turned round. Besides higher mass transport coefficients, these areas are also responsible for the formation of smaller bubble and droplet diameters than with the static mixer sieve plates. In contrast, when the sieve plates are used, only laminar flow patterns are generated at a higher volume rate within the channels of the sieves, but the efficiency is not enough for formation of turbulence, which is very important for high mass transport rates. Since the static mixer double axe was more efficient for mixing in the multiphase system without a significant pressure drop, it was used as standard for the following kinetic experiments.

3.2.2
Mass Transport Limitation at the G/L-Interphase

Besides the efficiency of mixing, the rate of hydrogen uptake into the liquid phase can be influenced by the hydrogen volume hold-up and, hence, hydrogen volume rate. Provided that the reaction is limited by mass transport at the G/L-interphase, the overall reaction rate should correlate with the volume rate of hydrogen, since the specific surface area increases with increasing hydrogen volume hold-up. This effect could be shown by photographic imaging for the biphasic system H_2/H_2O (Fig. 13). Another point that has to be mentioned is the increase of the mean bubble size and the broadening of the bubble size distribution as the hydrogen flow rate increases. The flow pattern changes from a homogeneous bubbly flow to an inhomogeneous churn turbulent flow regime.

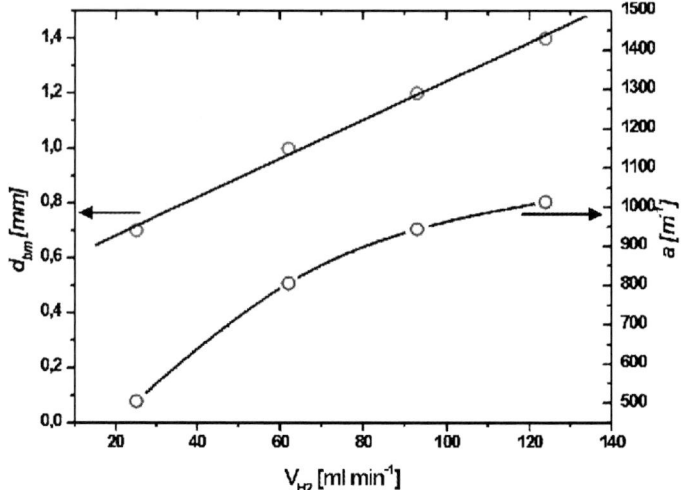

Fig. 13 Mean bubble sizes and specific surface areas of the dispersed gas phase in the biphasic system H_2/H_2O as a function of the hydrogen volume rate

In order to analyse the effect of increasing hydrogen volume hold-up and hence G/L-specific surface area on the overall reaction rate and to evaluate the mass transport rate at the G/L-interphase, citral hydrogenation was investigated at different hydrogen volume rates (Fig. 14). The selectivity to the isomeric unsaturated alcohols nerol and geraniol did not change, exhibiting values of 51–53% for nerol and 47–49% for geraniol.

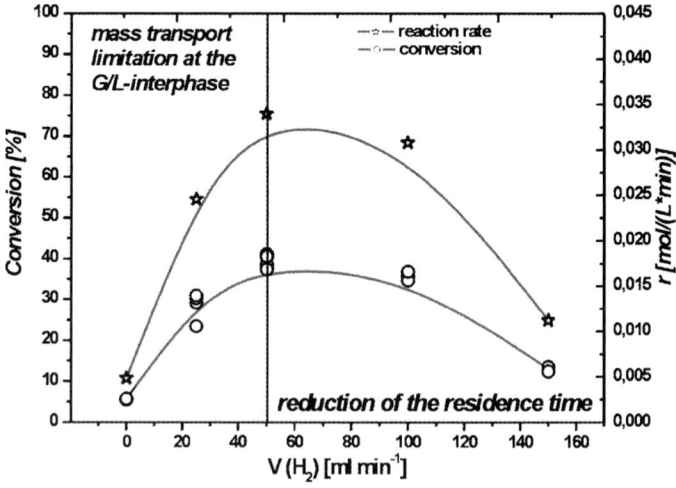

Fig. 14 Influence of the hydrogen volume rate on the observed reaction rate; $T = 60\,°C$, $p_{H2} = 5$ bar, $V_{aq} = 300$ mL min^{-1}, $V_{org} = 50$ mL min^{-1}, $V_{Feed} = 3.0$ mL min^{-1}

At low hydrogen volume rates, there is a direct correlation between the overall reaction rate and the hydrogen volume hold-up until a maximum point for the reaction rate is reached at $50-70$ mL min^{-1}. Increasing the hydrogen volume rate further results in a decrease of the overall reaction rate, which is presumably due to decrease of the organic phase residence time. Citral hydrogenation in the loop reactor seems to be limited by mass transport at the G/L-interphase at lower hydrogen volume rates ($0-70$ mL min^{-1}). Therefore, all the following kinetic experiments were performed at a volume rate of at least 100 mL min^{-1} in order to be sure that the overall reaction rate is not influenced by the rate of hydrogen uptake into the liquid phase.

3.2.3
Mass Transport Limitation at the L/L-Interphase

An important point that is commonly discussed in L/L-biphasic catalysis is the location of reaction. One theory is that the reaction takes place within the bulk of the catalyst phase, whereas the other theory demands the reaction to be located at the phase boundary. The discussion is not only interesting from an academic point of view but is also very important for reaction engineering and kinetic modelling, as shown by Cornils et al. [9] in the hydroformylation of propylene. In the case of substrates that exhibit a very low solubility in water, the question then arises about the degree to which mass transport at the L/L-phase boundary can be optimized by hydrodynamics. If the reaction takes place at the phase boundary, the most important parameter in order to accelerate mass transport would be the specific surface area. This can be increased for example by increasing the volume rate of the organic phase, as shown for the system H_2/H_2O. On the other hand, if the location of reaction is the bulk of the catalyst phase (according to the classical view based on the film model), the mass transport coefficient k_{LL} would have to be maximized in order to get a higher degree of catalyst utilization and hence a better space–time yield. Since the mass transport coefficient is a function of the Sherwood number (Sh) and this directly correlates with the Re number (Eq. 12), mass transport coefficients can be increased by a higher catalyst phase volume rate. The effect can also be attributed to a decrease of the length of the laminar boundary layer, since the mass transport coefficient is also a function of this parameter (Eq. 13):

$$k_{LL} = \frac{Sh \cdot D_{aq}}{\overline{d}_D}$$

$$Sh = f(Re, Sc) \tag{12}$$

$$k_{LL} = \frac{D_{aq}}{\delta_{aq}} . \tag{13}$$

The location of reaction can be evaluated according to the results published by Chaudhari and Delmas [16]. Thereby small amounts of unsulfonated triphenylphosphane, which is exclusively soluble in the organic phase, are added to the biphasic system (TPP : TPPTS = 1 : 100). Due to the interaction occurring between TPP and Ru, the water-soluble complex is attracted to the L/L-interphase, shifting a hypothetic reaction location within the bulk of the catalyst phase to the interphase. As a matter of fact, mass transport of substrates is reduced to diffusion of the substrates to the interphase. If the reaction normally takes place in the bulk of the catalyst phase, with a limitation of the overall reaction rate by mass transport at the L/L-interphase, the addition of TPP would result in an enhancement of the observed reaction rate. We observed an increase of the overall reaction rate in citral hydrogenation by a factor of almost two when TPP was added to the organic phase, giving rise to the conclusion that the reaction originally takes place in the bulk of the catalyst phase and mass transport limitation at the L/L-interphase has to be taken into account. Based on UV/VIS analysis of the organic phase during reaction, leaching of Ru into the organic phase and catalysis by Ru(II)-TPP in the organic phase could be neglected.

As already shown by Wiese et al. [17] mass transport rates in biphasic catalysis can be dramatically influenced by hydrodynamics in a tube reactor with Sulzer packings. Above all, the volume rate of the catalyst phase in which the substrates are transported by diffusion plays a decisive role in accelerating the mass transport rate. This effect was also investigated for citral hydrogenation in the loop reactor. Overall reaction rates and conversions as a function of the catalyst volume rate can be seen in Fig. 15.

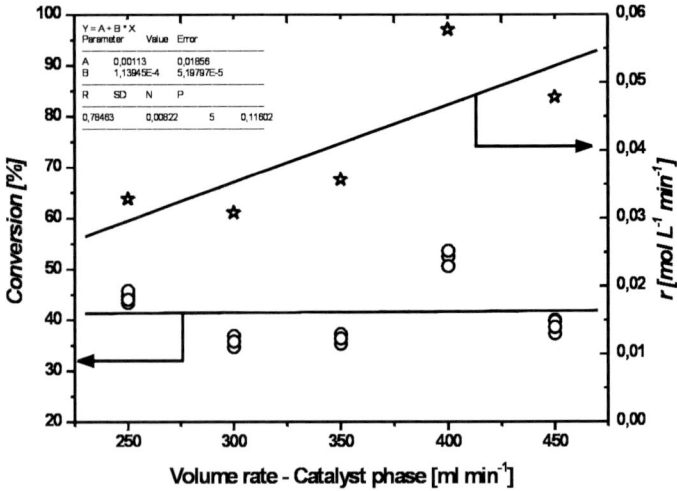

Fig. 15 Influence of the catalyst phase volume rate on the overall reaction rate; $T = 60\,°C$, $p_{H2} = 5$ bar, $V_{org} = 50$ mL min^{-1}, $V_{H2} = 100$ mL min^{-1}, $V_{Feed} = 3.0$ mL min^{-1}

As expected, the overall reaction rate increases with increasing catalyst volume rate. The effect can be explained in terms of the dependency of the mass transport coefficient k_{LL} on the Re number (Eq. 12). Due to the increase of the volume hold-up of the aqueous phase, the residence time of the organic phase decreases, so that the observed conversion degrees do not change within the limits of the investigated regime.

3.2.4
Kinetic Experiments and Reaction Mechanism

3.2.4.1
Hydrogen Partial Pressure

According to kinetic experiments performed in the batch reactor, hydrogen partial pressure only exhibits a reaction order close to zero, which is due to the fact that the amount of hydrogen dissolved in the catalyst phase is not the limiting factor. In order to check whether the experimental setup has an effect on the reaction order of hydrogen, similar experiments were also carried out in the loop reactor for citral hydrogenation (Fig. 16). Thereby, the hydrogen volume rate was always kept in the kinetic regime.

Increasing the hydrogen partial pressure initially causes an acceleration of the overall reaction rate until a plateau is reached beyond 7.5 bar. Further increase of the hydrogen partial pressure does not affect the overall reaction rate, which is in accordance with our results obtained in the batch reactor.

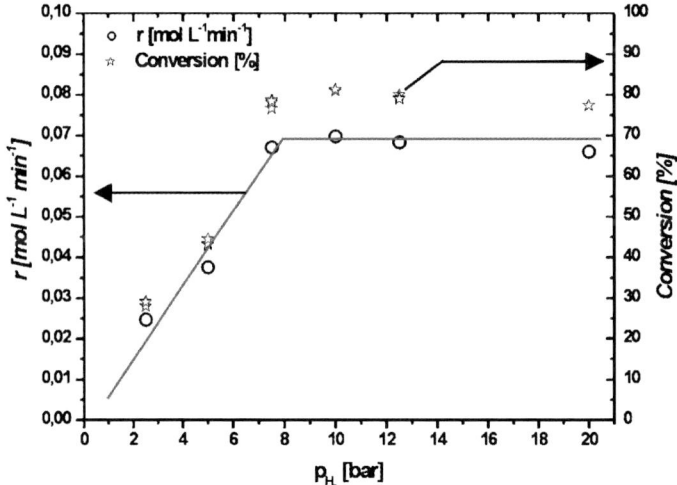

Fig. 16 Influence of the hydrogen partial pressure on the overall reaction rate; $T = 60\,^{\circ}\mathrm{C}$, $p_{H2} = 2.5\text{--}20\,\mathrm{bar}$, $V_{aq} = 300\,\mathrm{mL\,min^{-1}}$, $V_{org} = 50\,\mathrm{mL\,min^{-1}}$, $V_{H2} = 25\text{--}150\,\mathrm{mL\,min^{-1}}$, $V_{Feed} = 3.0\,\mathrm{mL\,min^{-1}}$

The initial increase of the reaction rate is possibly due to the rate of catalyst generation, which involves hydrogen as reducing agent.

3.2.4.2
Citral Concentration

Like the partial pressure of hydrogen, the citral concentration in the organic phase seems not to affect the reaction rate when carried out in the batch reactor. In contrast it has a significant influence on the reaction rate in the loop reactor, which is presumably due to the mass transport limitation at the L/L-interphase (Fig. 17). We calculated a reaction order of one for the citral concentration in the organic phase. By increasing the citral concentration in the organic phase the concentration gradient at the L/L-interphase is enhanced, giving rise to an increase of the mass transport rate and hence the overall reaction rate.

The reaction order of one is also in good accordance with the film theory, where the rate of mass transport linearly correlates with the equilibrium concentration of citral in the aqueous phase. As a matter of fact, the mass transport rate is of first order regarding the substrate concentration in the organic phase. Therefore, what is measured is in fact the rate of mass transport and not the rate of chemical reaction. This result is in our opinion a good example of how kinetic parameters could be falsified when the reaction is limited by mass transport and not kinetics.

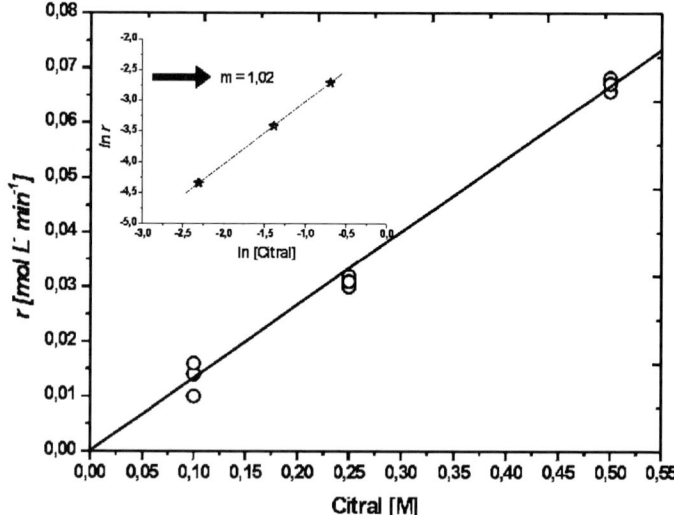

Fig. 17 Kinetic investigations on the influence of citral concentration on the overall reaction rate; $T = 60\,^\circ\mathrm{C}$, $p_{H2} = 5$ bar, $V_{aq} = 300\,\mathrm{mL\,min^{-1}}$, $V_{org} = 50\,\mathrm{mL\,min^{-1}}$, $V_{H2} = 100\,\mathrm{mL\,min^{-1}}$, $V_{Feed} = 3.0\,\mathrm{mL\,min^{-1}}$

3.2.4.3
Ruthenium Concentration

Since the overall reaction rate in the loop reactor is limited by mass transport at the phase boundary, one would expect that the Ru concentration has a weaker influence on the rate of reaction than in the batch reactor. We have carried out experiments at a Ru concentration of 0.005 M as well as at 0.01 M and observed nearly a doubling of the overall reaction rate, giving rise to a reaction order of 0.96 with regard to Ru. The result is somehow surprising, since it can be explained only in terms of a kinetic control of the reaction, like in the batch reactor. On the other hand, previous experiments clearly indicate a mass transport limitation at the L/L-interphase. So the question which arises is: how it can be possible that a multiphase reaction system is limited by both mass transport and kinetics?

A similar observation was also reported for the hydroformylation of propylene in a tube reactor by Wiese et al. [17], which was explained in terms of the following correlations. If the Ha number is very small (< 0.3), the overall reaction rate is limited by kinetics and the reaction predominantly takes place within the bulk of the catalyst phase. As soon as the intrinsic reaction rate gets faster, so that the Ha number has a value of 0.3 < Ha < 3, the reaction also takes place in the boundary layer to a certain degree. The rate of diffusion then has an important influence on the overall reaction rate. In the case of Ha numbers higher than 3, the reaction fully takes place within the boundary layer and the concentration of the substrate in the catalyst phase is zero. As a matter of fact, at Ha numbers between 0.3 and 3 there must be a transition regime where the overall reaction is limited by both kinetics and mass transport. Thus, the kinetic regime and also the location of reaction can be influenced by kinetic or hydrodynamic parameters. By increasing the Ru concentration, the intrinsic reaction rate is accelerated giving rise to an increase in the Ha number. The location of reaction is then shifted towards the phase boundary. Since the local citral concentration is higher in the boundary layer than in the bulk of the catalyst phase, an increase of the overall reaction rate is observed. In contrast, by increasing the rate of mass transport, for example by a higher catalyst volume rate, the location of reaction is shifted towards the bulk of the catalyst phase. Due to a decrease of the length of the laminar boundary layer the Ha number decreases and the overall reaction rate tends to be controlled by kinetics. Consequently, the degree of catalyst phase utilization, as well as the space–time yields, increase. The results in the loop reactor obtained at different kinetic and hydrodynamic conditions can therefore only be explained in terms of Ha numbers between 0.3 and 3, for which the overall reaction rate can be both limited by kinetics and mass transport.

4
Summary

Our investigations on reaction engineering aspects in aqueous multiphase catalysis focus on a better understanding of the interplay between mass transport and reaction. The regioselective hydrogenation of α,β-unsaturated aldehydes to the corresponding alcohols based on a water-soluble Ru(II)-TPPTS complex catalyst was thereby chosen as a model reaction. Reaction conditions were first optimized in a batch reactor and the general transferability of the catalyst system on all kinds of α,β-unsaturated aldehydes besides acrolein was shown. Recycling experiments with prenal and citral clearly demonstrated that the stability of the catalyst phase is primarily a function of the water solubility of the product, leading to a fast deactivation of the catalyst phase in the case of an accumulation of products in the aqueous phase. After an evaluation of the mass transport rates at the G/L- as well as the L/L-interphase by an experimental and theoretical approach, the reaction rate was modelled based on kinetic results by means of an optimization routine. Furthermore, kinetic experiments were also performed in a loop reactor. For the first time, we investigated the effect of hydrodynamics on the rate of mass transport in an aqueous multiphase system. An important parameter influencing the overall reaction rate is the geometry of the mixer, which determines the specific surface area and the degree of turbulence within the continuous catalyst phase. The influence of the ratio of volume hold-ups and volumetric flow rates on the mass transport rates at the G/L- and L/L-interphase was then analysed. Kinetic experiments revealed that the overall reaction rate was limited by kinetics and by mass transport at the L/L-interphase as well. Therefore, the kinetic regime as well as the location of reaction is determined by the kinetic and hydrodynamic conditions applied, giving rise to different Ha numbers.

References

1. Cornils B, Kuntz EG (1995) J Organomet Chem 502:177–186
2. Cornils B, Herrmann WA, Eckl RW (1997) J Mol Catal A Chem 116:27–33
3. Cornils B (1998) Org Proc Res Dev 2:121–127
4. Auch-Schwelk B, Kohlpaintner C (2001) ChiuZ 5:306–312
5. Cornils B (1999) Top Curr Chem 206:133–152
6. Herrmann WA, Kohlpaintner CW (1993) Angew Chemie 105:1588–1609
7. Chaudhari RV, Seayad A, Jayasree S (2001) Catal Today 66:371–380
8. Önal Y, Baerns M, Claus P (2004) Fundamentals of biphasic reactions in water. In: Cornils B, Herrmann WA (eds) Aqueous biphasic catalysis. Wiley, Weinheim
9. Wachsen O, Himmler K, Cornils B (1998) Catal Today 42:373–379
10. Grosselin JM, Mercier C, Allmang G, Grass F (1991) Organomet 10:2126
11. Joó F, Kovacs J, Benyei AC, Katho A (1998) Catal Today 42:441–448

12. Fache E, Mercier C, Pagnier N, Despeyroux B, Panster P (1993) J Mol Catal 79:117–131
13. Chaudhari RV, Gholap RV, Emig G, Hofmann H (1987) Can J Chem Eng 65:744–751
14. Mills PL, Chaudhari RV (1997) Catal Today 37:367–404
15. Okufi S, Perez de Ortiz ES, Sawistowski H (1990) Can J Chem Eng 68:400–406
16. Chaudhari RV, Bhanage BM, Desphande RM, Delmas H (1995) Nature 373:501–503
17. Wiese K-D, Möller O, Protzmann G, Trocha M (2003) Catal Today 79/80:97–103

Author Index Volumes 1–24

The volume numbers are printed in italics

Braga D (1999) Static and Dynamic Structures of Organometallic Molecules and Crystals.
   4: 47–68
Breit B (2007) Directed Rhodium-Catalyzed Hydroformylation of Alkenes. *24*: 145–168
Breuzard JAJ, Christ-Tommasino ML, Lemaire M (2005) Chiral Ureas and Thiroureas in
   Asymmetric Catalysis. *15*: 231–270
Brüggemann M, see Hoppe D (2003) *5*: 61–138
Bruneau C (2004) Ruthenium Vinylidenes and Allenylidenes in Catalysis. *11*: 125–153
Bruneau C, Dérien S, Dixneuf PH (2006) Cascade and Sequential Catalytic Transformations
   Initiated by Ruthenium Catalysts. *19*: 295–326
Brutchey RL, see Fujdala KL (2005) *16*: 69–115
Butler PA, Kräutler B (2006) Biological Organometallic Chemistry of B$_{12}$. *17*: 1–55

Candy J-P, Copéret C, Basset J-M (2005) Analogy between Surface and Molecular Organo-
   metallic Chemistry. *16*: 151–210
Castillón S, see Claver C (2006) *18*: 35–64
Catellani M (2005) Novel Methods of Aromatic Functionalization Using Palladium and
   Norbornene as a Unique Catalytic System. *14*: 21–54
Cavinato G, Toniolo L, Vavasori A (2006) Carbonylation of Ethene in Methanol Catalysed
   by Cationic Phosphine Complexes of Pd(II): from Polyketones to Monocarbonylated
   Products. *18*: 125–164
Chandler BD, Gilbertson JD (2006) Dendrimer-Encapsulated Bimetallic Nanoparticles: Syn-
   thesis, Characterization, and Applications to Homogeneous and Heterogeneous Cataly-
   sis. *20*: 97–120
Chatani N (2004) Selective Carbonylations with Ruthenium Catalysts. *11*: 173–195
Chatani N, see Kakiuchi F (2004) *11*: 45–79
Chaudret B (2005) Synthesis and Surface Reactivity of Organometallic Nanoparticles. *16*:
   233–259
Chlenov A, see Semmelhack MF (2004) *7*: 21–42
Chlenov A, see Semmelhack MF (2004) *7*: 43–70
Chinkov M, Marek I (2005) Stereoselective Synthesis of Dienyl Zirconocene Complexes. *10*:
   133–166
Christ-Tommasino ML, see Breuzard JAJ (2005) *15*: 231–270
Chuzel O, Riant O (2005) Sparteine as a Chiral Ligand for Asymmetric Catalysis. *15*:
   59–92
Ciriano MA, see Tejel C (2007) *22*: 97–124
Claus P, see Önal Y (2008) *23*: 163–191
Claver C, Diéguez M, Pàmies O, Castillón S (2006) Asymmetric Hydroformylation. *18*: 35–64
Clayden J (2003) Enantioselective Synthesis by Lithiation to Generate Planar or Axial Chi-
   rality. *5*: 251–286
Connon SJ, see Blechert S (2004) *11*: 93–124
Copéret C, see Candy J-P (2005) *16*: 151–210
Costa M, see Gabriele B (2006) *18*: 239–272
Cummings SA, Tunge JA, Norton JR (2005) Synthesis and Reactivity of Zirconaaziridines.
   10: 1–39

Damin A, see Bordiga S (2005) *16*: 37–68
Damin A, see Zecchina A (2005) *16*: 1–35
Daniel M-C, see Astruc D (2006) *20*: 121–148
Dechy-Cabaret O, see Kalck P (2006) *18*: 97–123

Subject Index